CW01645305

The

Southern Way

Keith Dawe

Eastleigh Enginemen

Special Issue 19

www.crecy.co.uk

ISBN 9781800352612

First published in 2022

Contact details
All editorial submissions to:
The Southern Way
Crécy Publishing Limited
1a Ringway Trading Estate
Shadowmoss Road
Manchester M22 5LL
enquiries@crecy.co.uk

A CIP record for this book is available from the British Library.

Publisher's note: Every effort has been made to identify and correctly attribute photographic credits. Any error that may have occurred is entirely unintentional.

Printed in India by Thomson Press

Crécy Publishing Limited
1a Ringway Trading Estate
Shadowmoss Road
Manchester M22 5LH

www.crecy.co.uk

Front cover:
No. 34034 *Honiton* cleaned and prepared outside the top of Eastleigh Shed ready to work an inter regional service. She was one of the few Light Pacifics to last through to July 1967.

Rear cover:
A very atmospheric view of the interior of Eastleigh shed.

Note: unless stated otherwise, all colour images are by Bob Winkworth

Contents

The
Southern Way
The regular volume for the Southern devotee
59

Preface

Looking back to a time in my life, now well in excess of fifty years ago, I recall the main question that my friends and contemporaries asked was 'Why'. Why would a successful grammar course student, destined for higher education, suddenly throw it all away to take on a calling of hard physical labour, heat and dirt when such a way of life was fast becoming unfashionable?

In my defence I can only say that I always had a very real interest in railways and particularly the steam locomotive. Not just as a casual line-side observer of their comings and goings, but more about how they actually worked. Indeed, as a child I made several drawings of various classes of steam locomotive to as near scale as my young talent would allow from published information.

Inevitably school life reaches the stage where decisions have to be made. What GCEs (as they were then known) was I to concentrate on and, in order for me to answer that, what did I intend to do as a career? Then, as now, not many fifteen-year-olds have even thought about their eventual career, never mind settled on a path.

So it was that in March 1964 my father asked this very question and I had absolutely no answer to give him, save the promise that I would think seriously about my future.

I naturally inclined toward the railways and the steam locomotive, but my father was a precision engineer and advised against engineering as he thought it a bit of a dying profession. Hence I thought I would enquire about locomotive driving.

So, without telling either parent, I wrote to the 'manager' at the engine shed at Eastleigh to enquire what GCEs they recommended I take to be able to pursue a career on the footplate. I had a very nice letter back from the Motive Power Department Shed Master, Eastleigh, informing me that GCEs were not necessary as they would give in-house training, subject to my being assessed as competent and medically sound. I was also invited to an interview the following week where I could have a chat to assess my suitability. I needed to bring my letter of invitation with me.

As it happened, the following week was the first week of the Easter holiday and so on the appointed day I told my mother that I was visiting friends and would be back for dinner. Instead I actually cycled from my home on the western side of Southampton to Eastleigh and made my way to the steam shed. Here I followed the instructions; up the dark-panelled staircase into the open plan office, where I was directed to another office at the far end.

The man who interviewed me in the big office at the far end was not who I later knew to be the shed master, so I am guessing he was a stand-in for some reason. He explained that you didn't just get a job as an engine driver, but took very many years of experience to achieve that post and had to become a fireman first and, before that, a cleaner. In any event, it was illegal to become a fireman before you were sixteen. However, if I became a cleaner now, I would be in a position to become a fireman by the time I was sixteen.

Having agreed that this was a good idea, a young lady appeared and showed me to another, smaller, office, which I later knew was the haunt of Mr Legg, the chief clerk.

I was set a comprehension test and given a series of basic maths problems to solve and told I had half an hour to complete. I have to say that I found the maths problems pretty basic and after ten minutes was at a loss to know what I should do next.

Driver Peter Bevan and fireman Keith Dawe, a former Tank Gang pairing reunited in later life as a guest crew at the Swanage Railway. This is in the preservation era, hence the hi-vis jackets. The engine is No. 34072 *257 Squadron*. *June Bevan*

Mr Legg returned to his office and suggested that I ought to go over to the medical officer and have a complete medical exam. The MO was situated in the Carriage Works in Barton Road, which involved me cycling there – again with a form to say why I was visiting.

Having had the medical examination by the doctor, I was taken outside, still in just underpants and socks on a not over warm day in late March and pointed in the direction of a building that I recognised as the top of the engine shed office, some half a mile distant. I was told to observe the semaphore signal attached to the top of the building and asked if it was 'off' or 'on'. I had no idea what this meant and just replied that it was 'down'.

The doctor then disappeared indoors, leaving me in the cold outdoors. Presently he returned and enquired 'now – off or on'.

'Well it's up now.'

'OK' and again he disappears, only to reappear half a minute later and enquire, 'What about now – off or on.'

'It's down again now.'

'Good, that will do. You can come in now.'

While still in a state of undress, I had to read a board of letters arranged in rows and of diminishing height with one eye covered with a bit of card and then a lantern with coloured slides that showed a red, green, blue and yellow filter.

That was it. I was instructed to get dressed and go home. I would be contacted shortly.

Shortly turned to be about three days before a letter arrived offering me a job as an engine cleaner. Starting pay was £4 10s for a forty-four-hour week, 8am to 4.30pm with a thirty-minute lunch break Monday to Friday, and 6am to 12 midday on Saturday. I could also start on Monday, 13 April 1964.

That was the easy part; now the hard part to persuade my parents to let me give up my education and, equally, did I actually want to give it up!

Having thought long and hard about it, I decided that this was what I wanted and surprisingly my parents agreed that if I really desired to pursue this career path they would not stand in my way. I understand that when they informed my school they were not so encouraging and considered it a very bad decision.

Even so, it was on that Monday morning, when my former classmates were returning to school after the Easter break, that I was on my way to my first day as an employee at British Railways Motive Power Depot, Eastleigh.

Was it the right move? Academically, of course, it was a complete disaster; but I still wouldn't have missed it for all the world.

It was a privilege to have served during those final and difficult few years of full steam operation on a busy main line and to be part of the transition to first-generation diesel and electric traction. More important was the sheer pride of being a fully trained steam engineman, one of the very last of many generations whose natural home was the footplate of a steam locomotive.

Of course, it was difficult at times, I was only a schoolboy who, due to circumstances, was thrust into a working environment that would normally be considered above the capabilities of such a youngster. But as you grew into the work, you either succeeded or you left the railway service. There was no halfway house.

Thankfully I succeeded, and in doing so experiencing the best years of my working life. It was a time of hard work with poorly maintained machinery, of camaraderie and laughter, and with a few tears thrown in as the boy was moulded into a man.

We lost our steam engines in July 1967 and embraced a new and exciting diesel/electric era – or so we thought. But we hadn't just lost our steam engines, we had lost the reason for having a railway.

The road haulage lobby was in the ascendancy and the railways were unable to compete with new working practices in industry, and of course we all had our own cars so we didn't need a railway. The railway I had known and loved had disappeared forever and in 1970 I decided to change my career direction and make a belated return to my studies. But the memories are still so vivid and so very dear to me.

Now the 'nipper', the mere boy, is a grandfather and an old man and whose former steam colleagues are getter fewer every year, I am persuaded to set down these memories in case they are of interest to the wider world.

Keith F. Dawe JP
Nursling, December 2021

34077

1
Eastleigh Shed
Setting the Scene

Originally, the locomotive facilities of the new London and Southampton Railway were situated opposite the terminal station at Southampton, however with the expansion of that station and its associated goods handling depot, the engine shed was moved about a half mile towards London, where a dedicated repair and servicing shed was constructed to serve the Southampton area together with the newly built docks.

Expansion in dock, freight and passenger traffic throughout the nineteenth century rendered the site at Northam inadequate for the growing number of engines that were needed to service this traffic.

Similarly, at the London terminal the locomotive works had become outdated and cramped. These two factors caused the Board of the then LSWR to purchase green land at Bishopstoke, which had become a major junction for north–south and east–west routes. The station called Bishopstoke was not actually at that village, but it was the nearest centre of habitation. The new complex of Carriage, Wagon and later the Locomotive works also had housing for the workforce that was expected to migrate from London. Thus a completely new township emerged from the fields around Eastley Farm, and henceforth the name of Eastleigh entered into the railway vocabulary. The large, fifteen-through-road engine shed was built adjacent to the works as part of this new complex and commenced operating in 1903. Consequently, after a lifespan of over sixty years, Northam shed closed and all locomotives and personnel transferred to Eastleigh. The site of Northam became a freight yard for that part of Southampton and the memory of the large and vital engine shed faded.

Contemporaneously with the development at Southampton, things were happening further up the line near the village of Bishopstoke. In 1841, a junction was created with a branch off the main line towards Gosport and a ferry crossing over Portsmouth Harbour to connect with that important naval port and town. Further expansion occurred in 1847 with a branch to Salisbury from the same point near to Bishopstoke and close to an estate called Eastley. The station was originally called Bishopstoke and later Bishopstoke Junction. It eventually became Eastleigh and Bishopstoke and finally just Eastleigh in 1923.

Above: **Packet 26 *Lamport and Holt Line* at the rear of Eastleigh.**

Opposite: **No. 34077 on shed at Eastleigh. Note also the removable wooden walkway across all the shed roads at the midway point.**

At this important north–south and east–west junction, locomotive servicing facilities were provided together with a two-road engine shed and turntable. I do not know when the first shed at Bishopstoke was built, but one certainly appears on the 1886 map as being parallel to the Gosport branch in the fork of the Gosport branch and the main line to Southampton.

According to another map of 1896, the shed had been rebuilt and was then orientated to be parallel with the main line to Southampton, still only a two-road shed and turntable and still in the fork with the Gosport branch. It was located in the present day as being between the railway junction and the Campbell Road bridge. It would have been completely removed when the new shed complex was built and opened in 1903.

Eastleigh was designed as a very large and important shed, arguably the most important on the LSWR system, later to become the Southern Railway and then British Railways Southern Region. Its designation was for mixed traffic, which meant that the traffic it provided power for encompassed the whole spectrum from local passenger, pick-up goods through to heavy, long-distance freight and fast passenger, and the all-important dock traffic.

Along with the ocean liner passenger traffic, vast tonnages of freight originated from or were taken to the docks. Imports included general merchandise, which usually had to be taken to the marshalling yards at Eastleigh to be sorted into the regular long-distance freights, perishable commodities that would invariably be taken directly to the London markets at Nine Elms and bullion from the Cape boats, which were always guaranteed a fast passage directly to Waterloo amid maximum security.

Eastleigh's allocation varied with time, but was generally around 100–110 locomotives of all types and descriptions from small tank shunting engines to large passenger engines.

Due to the adjacent locomotive works, the shed was a transfer point for machines coming and going for heavy repair and overhaul, thus along with the regular inhabitants many visitors could also be seen.

The main entrance to the shed was from the main line just south of Eastleigh station and controlled by Eastleigh West signal box. As this was a very busy shed, the exit and entry lines were separated to facilitate operations. Engines going off shed did not interfere with engines coming on shed and being serviced.

From the reception line an incoming engine was first required to stop at a little wood hut where the occupant booked the engine in and obtained instructions from the running foreman's office as to whether the engine was to have its fire cleaned for later use or have the fire dropped ready for routine boiler cleaning and service. Instructions were also given regarding which way round the engine needed to be for its next duty. During the later period of steam operation, it was also at this point that a blowdown valve at the bottom of the boiler was opened. The action of opening this valve was to clear scale and salts that collected at the bottom of the boiler barrel, which if left would decrease the lifespan and

A pair of 76s undergoing maintenance at Eastleigh. Notice the brazier ready for winter use.

efficiency of that boiler. This operation meant that the fireman had to have a reasonable amount of water in the boiler to start with and was a factor that had to be taken into account when the engine was being prepared for shed.

As the blowdown was the job of the man in the wooden hut, there was usually some friction between him and the fireman as to just how much water and steam really needed to be ejected, bearing in mind that it was the fireman who would have to clean the fire and leave his engine with enough water and steam to allow it to be safely moved around the shed.

From the blowdown hut the engine either passed the turntable, or used the turntable as appropriate, past the large office block with the massive water tank on top, and on to the pit road for fire cleaning. The driver would usually vacate his charge by the office and leave the engine to his fireman to move as required. The fire cleaning pit roads were alongside a double-sided coaling stage, where at peak periods up to six engines could be coaled simultaneously.

The fireman would rake all the clean fire back to the rear of the firebox using a long-handled clinker shovel before using the same tool to literally scoop out all the red hot clinker and dirty fire from the front. Having cleaned out the front, the clean fire would then be pushed from the rear to the front of the box and similar cleaning done to the rear and fresh coal applied to the clean fire, now at the front. All the red-hot clinker and char were discharged out of the side of the cab on the ground directly beneath.

The cab would be swept clean again before the fireman took his shovel to the front of the engine and, standing on the front frames, known as the bogie table, he would open the smokebox door and shovel out all the char that had built up. That done and the char swept from the frames, the next job was to enter the pit underneath the engine and, with a long rake, clear all the ash from the ashpan. It is a peculiar fact that no matter how careful you were to ensure you were upwind, the ash would always blow back and cover you in a delightful shade of pinkish white dust. After a good three-quarters of an hour of hot and dirty work, the fireman could leave the engine in the care of the shed engineman and go for a wash and brush up.

More modern engines were equipped with a rocking grate, which allowed half the grate area to be rotated through 90 degrees so that the dirty fire could drop or be pushed down into the opened ashpan.

This was a huge saving in the fireman's time and efforts as he no longer had to manually throw the fire over the side of the cab.

All engines left on the pit road came under the control and care of the duty shed engineman, who was responsible for filling the tank or tender with water and moving the engine under the chute for coaling.

Coaling was a very manual and laborious operation. In the middle of the coal stage was a single track full of loaded coal wagons. Shed labourers were employed to shovel out the coal from the wagons into skips or tubs, each holding one ton of coal. The wheeled tub had its own set of rails leading to a stop above and at right angles to the engine. When the tub hit the stop, the front was released and opened to allow the coal to cascade down into the engine's bunker or tender. A large engine took five or six tubs.

No. 34059 *Sir Archibald Sinclair* at the rear of the shed.

No. 34066 *Spitfire*, the Lewisham disaster engine, heads light toward Southampton West Docks, having just come off shed and reversed in the down platform.

During the last couple of years before the steam shed closed, the coal stage fell into disrepair and all coaling was done with the aid of an ancient steam shovel, until that too also gave up and a diesel grab was employed.

It should also be noted that the Southern main line to Weymouth and Salisbury was the last line in the country to fully work by steam. As other steam sheds in the north and west closed, the surplus coal stocks kept in emergency reserve at such depots were loaded on to wagons and dispatched south. The fuel loaded on to engines at Eastleigh during the final years was far from satisfactory and a major factor in poor steaming tendencies. By the time it had reached the fireman's shovel, it was nothing more than second-rate coal reduced to dust and small lumps, interspersed with a liberal amount of bricks, lumps of concrete, parts of trees and any other rubbish that happened to be in the way when it was scooped up. Suffice to say that the firemen at Eastleigh during this time probably worked ten times harder to maintain enough steam to run a busy service on a tight timetable than previously when good fuel was to be had.

The shed enginemen comprised two sets of driver and fireman, each working a three-shift system and whose duty it was to ensure that engines were dispatched from the pit road promptly to give room for continuous new arrivals.

The drivers had a list of the engines they were expecting and the next duty to which they were allocated, or whether they were to be stopped for maintenance. It was the driver's task to allocate the order in which the engines were to be stabled dependent upon their next departure time. To do this the engines had to be taken from the coaling stage, beyond the shed throat and reversed back up into the shed from the bottom end. From the throat the line extended to a triangle so that those engines that were too big to use the turntable could be turned by shunting round the three sides. One leg of the triangle also led to a secondary exit/entry to the shed that could be used if the main route was blocked. This exited to a junction on the Eastleigh to Botley line and was controlled by South Junction signal box.

The engine shed consisted of fifteen through roads and a short sixteenth road, accessible only from the top end, used for heavier repair work. It had heavy lifting facilities and other machinery not usually found in most sheds.

Engines departing left the shed from the top end. The middle road, No. 8, was always kept clear of engines as this was used as an escape route through the shed so that a one-way system through the whole complex could be maintained. Engines not booked to be berthed, i.e. only requiring coaling before returning, would use the middle road to exit the shed.

Shed roads Nos 1 to 4 were not used to berth engines in service, but were reserved for periodic maintenance. Every seven days a loco would be stopped for a 'shed day', where the boiler would be washed out and essential servicing carried out with repair work as necessary. At any time there would be around a dozen or so engines out of steam. This left ten shed roads available for berthing engines and each road held six engines under cover plus many more outside.

Shed enginemen comprised of very senior drivers who had elected to retire from main-line work due to health reasons. They spent the last few years of their working life on far less energetic and stressful employment. Their firemen were, conversely, young newly qualified men at the start of their career gaining valuable experience before venturing out onto the responsibility of working a train.

The duty shed enginemen have borrowed an available Standard 5 to shunt a couple of Bulleids ready for having their fires relit.

A pair of filthy dirty Bulleids at the rear of Eastleigh shed in the final months of steam operation.

No. 34071 having its fire cleaned on the 'big engine side' of the coal stage. The other side, which was accessible via the turntable, was known as the 'tank engine side'. Notice the dirty fire and clinker dropping down out of the ashpan and onto the ground as it is raked through the opened fire grate from inside the cab.

No. 34013 *Okehampton* waiting her next turn outside the top of the shed – probably No. 16 road.

Theirs was not usually a happy match, with the elderly, vastly experienced and usually contrite member of that elite section of railway life, the preserve of enginemen, coupled with young, inexperienced upstarts who knew everything and had everything to lean, with a zest and eagerness that only such youngsters possess.

Apart from moving engines into the shed, they were responsible for moving engines up their allotted berthing road as the lead engines departed. Engines stopped and out of steam for service would also require moving and for this purpose a small tank engine would be allocated as shed engine. Peripheral to these activities, coal wagons had to be shunted to extract the empty and replace with full ones. The large stores attached to the shed also needed stores wagons moved to unload vital supplies. Occasionally a withdrawn engine would need to be taken over to the scrap road and dumped awaiting disposal. During the mid-1960s the scrap road was lamentably full of machines that had served long and faithfully, but had no place in the modernisation programme.

Steam raisers were employed to light the fires on engines coming back from servicing and also to keep a check on the state of engines berthed ready for their next duty. These men were qualified firemen who had come off the main line due to medical reasons before they had made the transition to driver. They were just a part of the large workforce permanently employed within the confines of the shed and who were not a part of the footplate line of promotion; these including fitters of all descriptions, boilersmiths, coppersmiths, storemen and labourers.

Permanently attached to the shed staff, but in the footplate line of promotion, was a group of irrepressible lads, the very bane of all other staff and the most junior members of that band of brothers called enginemen. These were the larking, frolicking, devil may care cleaner lads learning the very basics of their career.

On the north side of the shed, adjacent to No. 1 road, were workshops and mess rooms for the highly skilled fitters and boilersmiths. On the opposite side of the shed adjacent to No. 15 road was the cavernous stores within which everything replaceable on a steam engine, from boiler water gauge glass tubes to engine cloths were exchangeable. A veritable treasure trove of goodies, and irresistible to cleaner lads who might be able to sneak in unnoticed. Next to the stores were the mess rooms for other shed staff. The sand furnace was situated at the far end and, in between, the dreaded cleaners' cabin.

This last-named den was very much akin to a large prison cell. The bare brick walls were painted in standard Southern gloss green to its mid-point and Southern gloss cream from there to the ceiling, some 10ft above the floor. The heavy reinforced door gave entry to a room some 10ft by 20ft containing a coal fired stove in one corner, battered lockers along one wall and a heavy cast iron and steel table bolted securely to the floor, along either side of which were heavy bench seats also bolted to the floor. The large window opposite the door had a wire mesh backed by heavy iron bars to prevent the ejection of objects through the window. A very large and substantially constructed stoneware sink with a cold water tap completed the furniture.

No. 80152 in the process of being turned on the table. The engine's vacuum pipe is connected to the turntable and uses the vacuum created for the train brake by the engine to operate the table mechanism. The turntable operator is also the operator of the blow-down point at the shed entrance.

Every shed road had access to a water column at the top of the shed so that engines could top up their tank before departing off shed. This is one such column and a Bulleid Pacific has just topped up her tender. Notice the fire devil next to the column to combat the worst of winter and stop the whole thing from freezing up.

Routine maintenance in shed roads 2 and 3. The person walking across the picture is the boiler tube cleaner, who operated a compressed air lance that cleaned out the multiple small fire tubes running through the boiler. He is wearing a mouth mask to combat the dust and grit of his job.

A busy scene at Eastleigh with the up Belle passing a down Bournemouth service. On the shed departure road wait three engines, the front one being No. 34066 *Spitfire*, waiting to go light engine to Southampton West Docks. She will be followed off shed by a 76 Standard and a Crompton diesel. If you look carefully over the top of the third carriage up in this view, you will just see the top of a grey box with a white cross. Inside this box was the route indicator clock face that the firemen of departing engines would use to inform the signalman at West Box of their intended destination as a light engine.

These spartan surroundings were the base and mess room of the cleaner lads and had evolved through successive generations as a counter to the worst excesses of that happy, hard-working and hard-playing group.

The stove was always lit, summer and winter alike, as the only source of hot water and heat. Each week one of the cleaners would be tasked with keeping the fire going and God help the lad who let the fire go out! The flue pipe from the stove rose vertically to the ceiling and inscribed upon this flue were several chalked lines with dates. During the winter months it was the practice of the duty stove attendant to test his skill and see just how hot he could get the fire. The whole stove would be glowing red hot and the test was to see how far up the flue he could also make to glow red. In these snug and warm surroundings during the lunch break, a number of wicked schemes were hatched to the detriment of some other section of the shed staff. Many an excursion on to the shed roof resulted in the fitter's stove chimney being inexplicably blocked, causing the smoke to fill the mess room at tea break. Or the old detonator was placed in an empty cigarette packet and casually thrown away into the storeman's fire as the perpetrator retreated to a safe distance. The resulting explosion would rattle quite a few windows. The detonator placed under the wheel of the shed engine to explode as soon as the engine moved would frighten old Jimmy Thomas, the driver, who had no love of cleaner boys even when they were being good.

Engine shed heating system 1960s style. Effective up to about 10ft!

The purpose of young lads of age fifteen years upwards starting their footplate career as cleaners was to educate them in the function and location of the machinery they would one day have to master. The reasoning was that if you cleaned it thoroughly, you had a pretty good idea what it looked like.

Cleaning railway engines is not achieved with soap and water; it is done with a mixture of rape oil and paraffin, and then brick dust and tallow for the bright parts. Using an old engine cloth, the oil and paraffin mix was rubbed into the accumulated grime until the paintwork appeared, which was then buffed up with a dry engine cloth to give a lustre that only oil can achieve. Bare metal parts were cleaned off with the oily mix before applying a paste of brick dust and oil to scour the metal. The top of the boiler also had to receive attention and with a large engine this could only be reached by standing on the handrail while balancing with one hand on the hot boiler cladding. No one had ever heard of Health and Safety at Work in the 1960s.

When the maintenance staff required a smaller person to assist in less accessible places, like fireboxes and water tanks, it was the practice to second a cleaner lad for the duration. All was valuable insight into the workings of a steam engine from both inside and out; vital to his future career on the footplate.

Steam engines used sand, dropped in front under the driving wheels, to assist adhesion in slippery conditions. Dry sand will only run if it is kept completely dry. For this reason sand boxes on the engine are usually sited as close to the boiler as practicable to keep them dry and moisture free. It therefore follows that the sand has to be hot and dry before filling and to this end a sand furnace, consisting of a hopper full of sand over a continuously lit coal-fired oven, formed the workplace of a shed labourer and one of his tasks was to tend the oven. Firemen preparing their engines would draw sand in a scoop and top up the sand boxes before leaving the shed.

Another feature of a large shed such as Eastleigh was the allocation of a breakdown crane, which was always kept ready for action and usually stabled at the top of No. 4 road. In the event of an emergency, members of the breakdown gang drawn from fitters and maintenance staff crewed the crane and its associated train.

To the London side of the shed was a very large, rectangular, three-storey, brick office block surmounted by an equally large water tank that served as the roof of the building. Such a large tank was required to service the endless stream of steam locos coming on shed at all hours of the day and night. The tank was fed directly from water pumped up from the nearby River Itchen and although this water was filtered, a number of small baby fish did manage to be pumped in. These spent their whole lives within the tank and became large fish in time.

During the later life of the shed, the building's top floor was used as a store for paperwork. When constructed this floor was designed as a locoman's dormitory for the use of enginemen on double home turns. When all engines became 'common user' as opposed to having designated drivers and fireman, engine crews could be relieved en route. This meant that the engine could work through to a final destination,

Eastleigh shed roads 1–3 at a busy period with various engines in for routine maintenance, boiler washout etc. Note the removable wooden walkway that went right across all fifteen roads at the mid-point. Most firemen would just jump across the pit, but the more elderly drivers would use the walkway.

whereas the crew could work out and back to their home shed within a working day and the need to provide dormitory facilities at distant sheds was negated.

The first floor of the office accommodated the actual offices where the chief clerk and his many assistants and clerical staff worked, along with the all-important shed master and his secretary. Access to the first floor was by means of a passageway and stair, totally enclosed and separate from any connection with the ground floor, save for discrete access to a small room that had a glass cashier's window into one of the rooms on the ground floor. This was the pay desk, where all pay was dispensed in buff envelopes to men on the footplate line of promotion. This office was kept locked and out of bounds to enginemen, and when not in use on paydays the cashier's widow was sealed with a polished wooden board.

During these later years, the ground floor was the domain of the enginemen and was divided into several rooms. I have read that the original shed layout had the drivers' and firemen's cabins over on the south side of the shed itself, however, certainly post-Second World War the enginemen's lobbies were on the ground floor of the office block and the south side of the shed was taken up with a very extensive store; the dreaded cleaners' cabin and the sand house already mentioned. The actual working plan of any such complex would naturally alter to suit current needs and safety requirements.

From my recollection there were five rooms, four of which were interconnected by an open corridor on the north side. This was the side facing the works and not the shed yard.

The first room was the running foreman's office and the signing on desk. The signing on point was actually a purpose-made window from the London end of the office with a sliding glass window. The office side of this window was the haunt of the checker, whose job it was to maintain a watchful eye on the comings and goings and to just make sure that the stated signing on time was an accurate interpretation of the clock face over the window. The 'outside' of this signing on window was in fact a substantial lean-to building in order for men signing to do so with a degree of comfort.

The running foreman was responsible for the safe and timely working of the shed. He made sure that each duty had an engine of the right power and route availability as prescribed in the engine diagrams and that the men allocated to that duty were signed on and doing their job. This also meant that any latecomers were covered by the men who were booked on duty as 'spare'. It would be the foreman's job – or more likely the foremen's assistant – to ask a driver or fireman booked on as spare to cover for a latecomer or someone who had gone sick.

These requests were usually taken as being part of the job – it is what a spare set of men were there for, but just occasionally, if the men received such a request to cover for a sick person and they had already been on duty for about five hours, then such a request sometimes would not receive the same joyous acceptance that the foreman's assistant would have liked!

As time progressed towards the day when steam was supposed to be eliminated from the Southern Region, so fewer and fewer engines were available. Maintenance was severely curbed as fitters were away for retraining or retired. The state of maintenance became a cause for concern and frankly engine crews had to struggle on with locomotives that were so poor that a decade earlier they would never have been allowed anywhere near the shed exit.

Locomotives that had come to the end of their boiler safety ticket were withdrawn from service as they could not be used until the boiler was overhauled in the works – and that was not going to happen.

Thus it was that the same level of service was expected from fewer and fewer engines and inevitably there were many occasions where an inferior form of motive power had to be allocated to prestigious duties. I well recall one example, which was the reaction of a Nine Elms crew who were expecting a Merchant Navy-class locomotive to be prepared and ready for them to take off shed, only to find they had nothing more substantial than a standard class five. Their exact words to the duty foreman will not be recorded here, suffice to say they were not at all happy.

On another occasion the engine for the Portsmouth portion of the down paper train from Eastleigh could not be covered by anything at all. Not a single serviceable main-line engine was to be had. Literally the only engines on shed were two USA-class tank engines (Yankie Tanks) that would normally be the works shunter and the shed engine. One was fitted with a vacuum brake, the other only had a steam brake. With the vacuum-braked engine as the train engine and the steam-braked engine as the pilot, this is what worked the down Pompey Papers one morning. Anyone who has ridden on a Yankie Tank doing more than about 15mph will have heartfelt sympathy for the two engine crews that day as they rattled and bounced their very short wheel-based charges as fast as their little wheels would carry them.

Next to the foreman's office was the roster room. In here were rows of green-painted, glass-fronted cabinets, within which every duty that Eastleigh men were responsible for was detailed, along with critical timings (time due off shed etc) and relevant engine duty numbers. There were also cabinets displaying special working notices, which gave details of speed restrictions out of the ordinary, single-line workings, changes to signalling etc. It was information with which engine crews would need to be made fully aware. It was the duty of the engine crew to make themselves familiar with special notices and make their own notes as necessary. They were also usually printed in a supplement of the working timetable that was issued to footplate crew.

Through the corridor into the next room was basically an empty room with wooden bench seats around the walls. On the shed side of this room was the cashier' window, as already described. When not used on pay days, this was a quiet space that men could use for a more serious chat with their colleagues, or just quiet reading of notices etc.

There then followed two interconnected mess rooms. Originally one was for drivers and the other for firemen, however during the latter years when drivers had lost their god-like status, both rooms were used by all. In fact, a fireman would naturally take his break with his driver.

Eastleigh coal stage before being run down. Urie S15 No. 30498 and Urie N15 No. 30747 *Elaine* await coaling before being taken to be berthed in the shed.

These rooms were a somewhat more rumbustious environment than the previous rooms, with much friendly banter and raucous laughter. This was where the topics of the day were discussed in detail or general merriment and card schools abound. Both rooms had a large coal-fired stove that was attended by anyone who happened to be next to it when more coal was needed. On both stoves were the ubiquitous heavy iron kettle – always on the boil, and always in demand for the constant stream of tea cans that need filling. Upon the wall in the first of these rooms and in pride of place was the oval, cast brass cab-side numberplate of a LSWR T3 that had once been a long-term Eastleigh engine.

Unlike certain other similar establishments, Eastleigh enginemen's mess room was kept relatively clean and tidy, even though the air was usually thick with tobacco smoke as the majority of these men would be smokers. These mess rooms were always busy with the comings and goings of enginemen, and they were also always warm and cosy and a welcome haven before or after a hard day's – or night's – work.

Part of the shed complex housed a firing school, where the finer points and academic side of the skill of firing a steam engine were expounded to those cleaner lads who had acquired sufficient knowledge to allow them to progress onto the next stage in their career.

Subject to passing out in both academic and practical exams relating to the complete operation of the steam locomotive and the rules that govern safe and responsible working of both train and engine, the cleaner qualified as a junior fireman and took his place in the rota as shed engineman. Woe betide any of his former cleaner colleagues who did not respect his newly elevated position.

As a young fireman, he would be paired with a very senior driver on shed enginemen duties. The shed engineman would often work with the maintenance staff to reposition engines and the younger man would take the opportunity to discuss the technical aspects of various pieces of equipment in less busy moments.

During winter months the shed could become a very cold place to work. There was no form of heating, apart from the engines themselves, and the doors either end of the through shed were certainly never designed to be air-tight. In the coldest weather, braziers were lit at each end of the shed to combat the worst of the icy cold drafts. Outside the top of the shed a line of water columns gave access to each of the roads for engines to top up their tanks before departure, and in cold weather these each had a well-lit brazier to stop them from freezing.

As for as the enginemen, other than shed enginemen, their home shed was the start and finish of each and every day. While 'on shed' they came under the supervision of the shed master – 'the Guv'nor' – but immediately they went 'off shed' they were their own masters.

It was their shed, their home, and the pride they had in their shed was second only to that of their company, which was still, traditionally at least, the London South Western.

Here were their colleagues and here was the base for an elitist camaraderie, which could be relied upon even in the most adverse conditions. Many a time a long off-duty engineman could be found discussing some technical failure or detail with the fitters just for his own enlightenment. If equipment failed him out on the road, he would want to know why.

It was often said that to an engineman, his home shed was his main home, the wife and family were a second home.

Such loyalty and pride naturally led to inter-shed rivalry between the major sheds at Nine Elms, Feltham, Eastleigh, Salisbury and Bournemouth. Eastleigh was, naturally, by far the best and most important, but essentially they were all LSWR in both origin and tradition.

Eastleigh coal stage, 'tank engine side', with M7 No. 30133 being coaled using the 1-ton tubs.

2
The Dodger and Other Pick-up Goods

As the 1950s turned into the '60s, it became increasingly difficult to recruit lads of the right calibre to start a career on the footplate. It was no longer fashionable to embark on an arduous and sometimes dirty calling with unsociable hours. Indeed, this became a contributory factor in the demise of the steam locomotive, which was in itself an inefficient, labour-intensive form of transport.

Irrespective of how glamorous it looked from the line side, it was a very tough way to earn your living and those who followed in that career were immensely proud of their calling. The morale and camaraderie that existed between steam locomen was unsurpassed and a factor in the remarkable achievements made in the final two years of operation with inferior fuel and machines that were barely fit to traverse the main line, never mind put up performances of the highest order.

The footplate line of promotion was continuous from junior engine cleaner to the most senior top-link driver. There were no short cuts and plenty of pitfalls. If you failed the grade at any point, your footplate career also failed. Safety and discipline were of the highest order, irrespective of the critical staff shortage. If a young fireman, thrust into responsible work with only a fraction of the experience his predecessors would have had, failed to come up to requirements either physically or mentally, he was removed from the footplate. It was something that happened not infrequently.

As the more senior driving posts were vacated through retirement, the upward shift in drivers and qualified firemen to drivers outstripped the supply of less senior firemen. Fortunately, by the late 1950s and early '60s a lot of the shunting turns were diesel operated and only required a driver. Similarly, many cross-country services were diesel multiple unit turns, again only requiring a driver. However, all the main-line services between London and Weymouth were still steam hauled, together with inter-regional services and all freight both local and longer distance.

Long-term Eastleigh resident No. 80151 at the rear of the shed. Notice in the background the jib of the ancient steam grab used after the coaling stage was abandoned. This also eventually failed and a diesel-tracked grab was employed.

As the number of serviceable locomotives decreased and the coal stage fell into disrepair, so coaling tended to be carried out in various parts of the shed.

Tank Gang regulars. Nos 80139 and 41319 on shed at Eastleigh.

The effect of this rationalisation of steam was that a competent fireman would quickly find himself working alongside more senior drivers on duties that demanded a very high level of skill. Clearly, if the driver was supplied with the means to do his job and the fireman worked the engine within a safe manner, it didn't really matter what age his fireman was. By and large most drivers accepted the situation, but a few of the more senior ex-LSWR men took exception to sharing their footplate with a mere 'boy' and made their views known; even to the extent of pointing to an imaginary line across the footplate with the stipulation that the fireman was to keep to his side of the engine and not to cross over to the driver's side.

Having spent time pottering around the engine shed on shed engineman's duties and becoming thoroughly familiar with the requirements of heavy machinery on the move, the most junior of firemen would be allocated to duties outside the boundary of his home shed. These consisted of those shunting yards that were still steam operated and included two Loco Works shunters, Carriage Works shunter, Pre-Assembly Depot shunter, Town Yard shunter and station pilot. The major marshalling yards had, by the early 1960s, been dieselised with single-man operation, but from time to time the diesel engine would fail or need maintenance and be temporarily replaced by a steam engine – complete with fireman. It was not always a welcome interlude by a driver used to the relative ease, comfort and warmth of a diesel shunter.

It is often told how enginemen would cook their breakfast on a shovel in the fire and the impression is that this happened while on the move. Nothing could be further from the truth, as the draft on the fire would make short work of anything placed in the firebox upon a shovel. But it is true that such culinary indulgences happened when the engine was stopped and the fire quelled during a break at some more distant country goods sidings. Nothing could quite compare with the pleasure of having a fried breakfast out in the open on a summer's morn when the sun was climbing up into the sky, the birds were singing in the trees, the engine was gently simmering away to itself and all was well with the world.

However, the first occasion on which the writer experienced such quaint goings on was a complete disaster. It happened that the diesel shunter at Eastleigh West yard was away sick for a couple of days and had to be replaced by a steam tank engine with, of course, a fireman – *me*. By great fortune the driver booked for the week was a gentleman who welcomed his young mate as a source of company, and who he regaled with tales of his younger days as a fireman. Of the old drivers he spoke of, most had passed on to greater things in that celestial engine shed. He taught his very green fireman many things about engines and how things used to be done in the old days, including the delights of frying breakfast on the shovel.

In order to demonstrate the technique and to relive his younger days just one more time before he retired, he announced that he would bring in the ingredients and on the morrow he would partake of a fry-up.

The following day he duly arrived armed with bacon, eggs, half a tomato, two sausages, a slice of bread and a knob of best lard, all to be cooked up during the break after the Salisbury Goods had arrived and been dealt with. On this particular morning the signalman informed the shunter, and the shunter informed the driver, that the Salisbury Goods was delayed, so shunter and driver agreed to take breakfast early and deal with the tardy Salisbury goods after the break.

So in the small hours of a spring morning, when the previous round of activity had come to a conclusion, the fire was allowed to die down a bit, the engine trundled into the engine siding, the boiler was filled and the handbrake was applied. All was peaceful and warm inside the cab.

'Right ho, young man,' said Arthur the driver, reaching up into the locker at the back of the cab. Taking out his newly acquired plastic box inscribed 'Tupperware' and placing it reverently on his seat, he instructed, 'Pass us yer shovel mate.' Taking a clean engine cloth from his pocket, he wiped the dust from the coal-polished blade. Balancing the shovel on the glowing fire immediately under the fire door, a knob of lard was placed at the base of that shovel blade and it soon obliged by becoming a sizzling liquid.

Sausages were arranged around the slice of bread within the molten fat, next were added the tomato and bacon, leaving just enough room for the egg to go in slightly later. 'Off yer pop, old son and make us a can of brew,' was the cheery request, and I left dear old Arthur bent over his sizzling breakfast, happily poking the sausages to make them cook all round.

Over in the shunters' cabin, where the ever-boiling kettle lived, things were not quite so peaceful. One of the shunters was on the telephone, evidently talking to the signalman at West Box. Replacing the receiver rather abruptly, he informed this recently entered fireman that he was given 'duff gen' and the Salisbury was coming through Chandler's Ford and would be here shortly.

Now, an experienced fireman would know that, given this changed information, there would still be a good twenty minutes before the train rumbled into the yard and stopped, the shunting engine brought out and any work commenced. However, this was not an experienced fireman, this was me, a very young fireman with much to learn. In his mind he knew that his engine did not have enough pressure in the boiler to be doing anything worthwhile, and to his mind things were required immediately. It was with this alarming news that he hastened back across the yard to where his engine reposed.

Now, there is a device on all steam engines called a 'blower'. Its function is to send a jet of steam up through the smokebox and out the chimney, thus creating a vacuum in the smokebox, which the laws of physics says is an unnatural state. On a steam engine, nature replaces this particular vacuum by drawing air through the fire tubes from the fire grate and thus draws up the fire by creating a forced draft.

On re-entering the peaceful, aroma-filled cab, the news of the impending arrival was passed to Arthur, who was still bent about his cooking, having now deposited his egg into the sizzling fat with the fire doors about three-quarters closed. At the same instant as the news was imparted, this young fireman, with one swift and practised move, fully whipped open the blower valve.

There was a rush of steam out of the chimney, generating a kind of whoosh from the firebox and there, right before the driver's eyes, bacon, egg, sausages, tomato and fried bread were all neatly plucked from the shovel and sucked into the fire.

For a moment, or two, dear Arthur stared at the now empty shovel in sheer disbelief, words utterly failing him. Standing up slowly, face red with heat and rage, he turned on the hapless fireman and pointing to the now empty shovel. Recovering his power of speech, he shouted, 'You silly little sod, that's my bleedin' breakfast yer've put up the chimney.'

3MT 2-6-2T on empties returning to Fawley refinery at Southampton Central.

Chandler's Ford with the morning pick-up freight. No. 75068 is deputising for the usual 4MT tank.

It was some time before Arthur was encountered again, in fact just before he retired. Thankfully he must have forgiven the naivety of, what was probably, his last young fireman, for as a gift he gave me a little notebook on failures and remedies on diesel locomotives, which he thought might be useful later in this young man's career. That cherished neat handwritten notebook remains as a memento of a gentleman from a forgotten way of life.

Early firing days in the shunting links involved long hours of too-ing and fro-ing behind the scenes and out of public sight. Local freight trains, known as pick-up goods, which plied between the various local goods yards, would arrive back at the Eastleigh marshalling yards to be sorted into long-distance freights according to their destination. London-bound wagons would be put into the Nine Elms goods, westbound wagons to the Westbury or Exeter or Severn Tunnel Junction trains. Eastbound went into the Chichester, while Portsmouth had its own service from Eastleigh, as did Salisbury. South Midlands traffic would go on the Oxford and all wagons for places further north would travel up on the Feltham for rerouting at that major yard. Local traffic for Eastleigh itself would be tripped across the main line into Eastleigh town yard for unloading. Conversely, the main long-distance freights arriving at Eastleigh needed to be broken down into local pick-up freights for their final destinations.

Throughout all this activity, the fledgling fireman mastered the art of boiler control and a thousand and one other skills that the aspiring engineman needed as second nature before he was allowed out on to the iron road though sun, rain and winter snow. This was where the playful youth also turned into a young man.

Most of his railway life so far had been away from the public eye, where appearance took second place and the chore of overall washing was not given the highest priority. However, one of the duties was that of station pilot, which involved pottering around the station area taking carriages from the carriage sidings into the station for trains starting from Eastleigh and putting carriages into the sidings from trains terminating at that station. There were also parcels vans to be attached or detached from trains stopping en route to further destinations and sometimes additional carriages attached to crowded trains.

It so happened that on one particular day when this young fireman was station pilot, a London semi-fast called and unloaded a mountain of mail sacks onto the platform. When that train had departed, it was the station pilot's task to take a parcels van into the platform to be loaded with the mail ready for attaching to another train going to a different destination. Engine and van were accordingly signalled into the platform, which was crowded with passengers, and duly stopped adjacent to the mailbags. While awaiting the platform staff to complete their work, this young fireman became aware that his driver,

Redbridge Sleeper Works' stone-built engine shed, designed to house the small tank engine allocated as works shunter. It was a cosy sort of place until it was 'modernised' by the resident shunting engine to form an open plan, light and airy space.

who was on the platform side, had female company. Presently two rather delightful-looking young ladies looked into the cab with a cheery 'Hi ya' and a wave. Any amorous thoughts were quickly dashed when the departing females were heard to remark to the driver, 'Your boy's a bit of a scruff though, in 'ee.'

Hence forth, overalls were washed regularly and the tatty shirt used as rags. Another lesson had been learnt. It was not just sufficient to be good at your work, you also had to look presentable and keep as clean as possible while doing it; you never knew when young ladies might pop their pretty heads over your cab door!

Henceforth boots were to be polished every day. The shiny-topped engineman's hat, known as a grease top, was kept clean and pinned down at each side in the fashion generally adopted by Eastleigh firemen. (Bournemouth men apparently favoured yanking the top down over one ear, while Nine Elms men seemed to have a passion for wearing a knotted handkerchief and no hat at all!) The black serge jacket was always hung up in the locker, not used as a cushion on the hard wooden seat, and you always kept a clean engine cloth in your pocket as well as the grubby one you were using.

Soon enough, time would come for you to move on up to the next link and the day dawned when you presented yourself at the appointed hour in the enginemen's lobby to meet your new, regular, driver for your first turn of duty in the Tank Gang.

By the early 1960s the work in this link consisted of local duties radiating from Eastleigh to places including Southampton, Romsey, Fareham, Winchester and Brockenhurst, mostly with tank engines, but also with smaller and medium-sized tender engines as well.

Some passenger work involving stopping trains to Southampton supplemented the more usual pick-up freights that were the staple of this group of duties. Local parcels trains also added interest and although this was all fairly mediocre work, for the young fireman it was still a quantum leap forward. He was now doing the work his heart was set upon. He was established with his own regular driver and everything they were required to do within that link, they did together as a recognised pairing.

The Romsey goods was a typical duty that entailed signing on at around 06.15 to prepare the engine, usually a large Standard class 4 tank universally called an 'eighty thousand tank' by footplate crews, and working the daily freight to Romsey. Before closure, the freight called at Chandler's Ford to pick up and set down wagons on the way to Romsey, which was reached at about 08.30. On arrival the train would be reversed into the yard headshunt to await the clearance of the following passenger train, which offloaded hordes of school children attending the Romsey schools from outlying areas. Once all was again clear, the goods pulled out of the yard into the platform and the engine was run around the train before spending about an hour shunting the up side yard. Empty wagons and newly loaded ones awaiting dispatch were put to one side while new arrivals were positioned ready for unloading. Coal wagons went on to the coal stathes, heavy loads under the yard crane and most others went into the goods shed.

The train from Eastleigh also frequently contained some wagons for the Admiralty establishment at Dean Hill. These special MoD wagons were painted a dark grey/green. The vans were padlocked and the open wagons securely sheeted down with tarpaulins stamped with the government arrows. On such occasions the brake van would be attached to the rear of the short train and when the line was clear of passenger traffic, they would be trotted the half dozen miles up to the Admiralty sidings.

On arrival the steel gate across the siding would be opened from within and the train engine only allowed inside to the arrivals road, where it was detached and run forward on to a short siding. A fussy little diesel shunter appeared to collect the wagons and disappeared with them behind high earthen banks, leaving the train engine to run forward again, ready to leave.

After a short wait the fussy diesel again came from its lair pushing wagons for dispatch, which were coupled on to the engine before the diesel fled back behind its barricade. For some strange reason the Admiralty had no wish to have a fire-breathing steam engine anywhere near its cordite plant, for this was an underground ammunitions dump.

A quick canter back to Romsey brought us back on the up side and more time would be spent shunting those sidings before heading off over the Nursling road towards the London main line at Redbridge, then on to the goods yard at Millbrook. Up to this point the engine was far bigger and more powerful than necessary for the work. Millbrook yard was a fairly big facility that served all the western side of Southampton, excluding the docks. Thus it was the railhead for quite a considerable amount of light engineering as well as household coal for the growing western suburbs and general merchandise. The yard had no shunting engine allocated and all such activities were carried out by the successive pick-up goods engines that called throughout the day. The added advantage was that all local freights could be held at Millbrook until a clear pathway became available through the busy Southampton section.

After spending a further hour or so sorting out Millbrook yard, the latest additions to the never-ending stream of loaded wagons for dispatch would be added to the shorter train brought in from Romsey and the whole ensemble of what was now some fifty to sixty wagons taken up to the marshalling yards at Eastleigh. This that required the full power of the engine.

A slightly later duty with the same type of engine was the Fareham goods, calling at Botley yard and Knowle siding as required before spending time organising the yards at Fareham. Fareham was an interesting place, being a four-way junction where the routes to Portsmouth, Gosport, Eastleigh and Swanwick converged, with much through cross-country traffic as well as local services.

Having finished with Fareham, the growing train would now reverse direction to proceed over the Swanwick route towards Southampton, calling in at Swanwick, Hamble and Netley before arrival at Northam, where the engine ran around its train to take it back across the main line and into Bevois yard. More shunting at Bevois preceded the jaunt up to Eastleigh.

Woolston, Northam, Southampton Town, Winchester City and Winchester Chesil yards were all served by other pick-up goods trains. Each of these goods trains were timed to leave Eastleigh yards during the morning, interlaced between the passenger services, and arrive back in an orderly succession during the afternoon and in good time to be resorted onto the longer-distance freights according to their final destination; the long-distance freights leaving the Eastleigh yards in the early hours of the morning.

Having shunted the yard at Netley, No. 75065 has pushed back into the up platform ready to cross over to the down line and continue its journey towards Fareham.

A cold morning at the rear of Eastleigh shed with Midland tank No. 41313, an old friend from the Tank Gang days. Treasured memories of a bygone world.

The first of these local goods to leave Eastleigh at around 05.30 was affectionately known as the Dodger and would work out to Totton via Northam and back again via Redbridge Sleeper works and Millbrook. The crew would spend the first three-quarters of an hour preparing their engine, invariably a small standard class 2 tank in the 41xxx series, known to Eastleigh men as Midland tanks, before collecting the train from East yard.

At the appointed time the ensemble of about twenty to thirty wagons commenced its journey down the main line to the first point of call at Northam yard. The Midland tanks were easy engines to work and had a spacious, weatherproof cab. Apart from a hooter instead of a proper whistle, the only peculiarity was a tendency for the engine to nose from side to side when taking steam and travelling at slower speeds. From the footplate, it gave the impression of waddling along the track like a duck.

On arrival at Northam, the engine uncoupled and ran forward into a siding to allow the resident diesel shunter to take off wagons for that yard and put on wagons destined for other yards on the route. While this activity took place, the fireman would seize the opportunity to make a can of tea in the shunters' cabin.

As mentioned at the start, the location of Northam yard was originally the site of the main engine shed for Southampton and the remaining buildings in use probably dated from the 1850s. Accommodation wise it was very typical of all shunters' cabins around the railway system; clean but not overtly so with cushions from long-scrapped coaches arranged on the otherwise hard wood benches either side of a large wood table. There was a population of resident rodents and as always the ubiquitous iron, coal-fired stove that was never allowed to go out, and a large, ever-boiling kettle in its place on top. As hot water was extracted it was always replaced with fresh, so at no time was it ever far away from boiling. Heaven help the person who failed to replenish that most vital of utensils. As it was never emptied, it is, of course, entirely possible that some of the scale formed from the water molecules in that kettle also dated from the 1850s!

No. 80152 on the tank engine side of the now disused coal stage. She would have had her fire cleaned and be awaiting the shed enginemen to take her over to the shed and coaling facilities. Notice the headcode discs. This particular headcode was etched into the heart of all Eastleigh men. Two on the left only meant one destination in this part of the country – 'Light engine for Eastleigh Loco', aka 'the going home boards'.

Beside the door was a row of coat hooks, all overloaded with piles of greatcoats, raincoats, waterproofs and sou'westers, while underneath were rows of various-sized boots. There were always more clothes than shunters that owned them and under the top layer were probably discarded items owned and worn by employees long since departed. There was also a desk of ancient lineage upon which resided a couple of equally ancient Bakelite telephones. One of these connected directly to the local signal box, Northam Junction, the other to the main railway network. Alongside these venerable instruments was the train log, kept to record the arrival and departure times of all traffic using the yard. Finally, there was a row of wooden lockers that had ceased being able to lock a few generations earlier; a stone sink with brass taps and a wood drainer. All were housed in a hut only just large enough.

Overall there was the feel and smell of warm snugness, a haven from the foulest of weather, which, on a steam-operated railway, rarely interrupted operations. It was here that successive generations of railwaymen had worked, taken their meals and, in quiet spells, played cards or slept upon a cushioned bench. Here too generations of firemen had called to fill their tea cans with the much-boiled water.

Tea made and the first cup drunk, the engine regained its place at the head of the train for the journey out to Totton yard. As with all smaller tank engines, water was always a potential problem. For that reason, the Dodger was one of the few goods trains that was actually booked to stop at Southampton Central station and the train would clank its way into No. 4 platform and squeal to a stop opposite the water column at the platform end. While driver and fireman were engaged in filling the tank – the driver turning the water on and the fireman on top of the tank holding the thick leather pipe into the filler hole – the early Portsmouth to Salisbury diesel passenger train would arrive and depart ahead of the goods.

With the tanks filled up, the Dodger restarted its trek, rattling, squealing and clanking its way past Millbrook and then Redbridge; over the old wooden viaduct – soon to be replaced and realigned with a new concrete structure – taking the line across the head of Southampton Water; round the 'S' bend through Totton station with its gated level crossing, and into Totton yard.

Leaving the train in the arrival road, the first job was to extract all the empty wagons from the coal road and replace them with those that had just arrived. This was always done first as the coal delivery firms would be waiting to bag up the mineral into hundredweight bags for loading on to their lorries and onward delivery to the countless homes then warmed with coal fires.

Totton yard was not a very interesting place, situated as it was away from the station, separated from the main line to Bournemouth by the single-track branch line to Fawley and bordered on either side by the untidy back yards of modest commercial premises. There was little to catch the attention of the relaxing fireman as his driver worked the engine up and down the yard shunting wagons hither and thither at the request of the porter/shunter, who had to walk down from the station to attend operations in the yard. Greetings were exchanged with fellow Eastleigh crews on passing trains, and the occasional necessity to attend fire or boiler were the only requirements until the work was complete and breakfast could be taken.

At Totton, breakfast was eaten in their shunters' cabin, largely because it was a comparatively new building and therefore cleanish. As the shunter had departed on his walk back to the station to resume his other duty as a porter, driver, fireman and guard could eat their food in peace and when they were ready, give a toot on the hooter to let the signalman know that they were ready to set off back towards Redbridge on the return journey.

Redbridge was a railway-owned complex situated beside the head of Southampton Water, where ships from Scandinavia unloaded cut sleepers of pinewood. These were then immersed in vats of creosote and let dry. Thus the rail traffic consisted of empty wagons in and loaded wagons of new sleepers out, destined for the permanent way Pre-assembly Depot at Eastleigh. There was also a small shunting engine permanently allocated to the equally small, but substantial, stone-built engine shed. The water column associated with the engine shed was the reason why the Dodger called at Redbridge on the return journey, rather than the operationally more convenient access to the yard on the outward journey. It was important to fill the tanks up before the trip back to Eastleigh.

The little engine shed at Redbridge was just big enough to take one small engine and was a snug sort of place, that is until one day when the tank engine entered the shed, failed to stop and exited by a slightly unorthodox route through the back wall. The exact thoughts of the unfortunate driver at this time are not recorded, but the rubble was cleared away and for the rest of its life the shed remained in its new configuration, open to the wind and rain sweeping up Southampton Water and very definitely no longer snug.

The homeward passage, with about thirty wagons in tow, would be interrupted by the obligatory call at Millbrook yard to engage in shunting, as required, and taking on further wagons awaiting departure. However, care was exercised in not adding too many wagons to the train as the following goods from Romsey had a much larger engine and would be able to take a heavier load.

Once a passage through the busy Southampton section, with its twin-track tunnel under the town centre, had been found, the Dodger would be signalled away from Millbrook, usually with gestures from the signalman indicting that there was other traffic following and he didn't expect the local goods to dawdle. Not that he needed to worry himself; this was the homeward-bound leg of a day that started in the early hours of the morning. There would be no question of dallying around and the little engine would be made to canter up the main line just as fast as she could manage with her train of wagons rattling and groaning behind her, in the manner of all goods wagons when they are made to hurry.

3
Bournemouth Link

All too quickly, time passed by and the days of local work gave way to the more demanding duties of the Bournemouth Link. The bidding adieu to that most amiable of drivers in the Tank Gang was tinged with trepidation as the new regular man was more experienced and with a reputation as a hard driver. His great delight was, seemingly, trying to make his fireman run short of steam. Many tales were related by disgruntled firemen of how driver David 'Thrasher' Batten had mercilessly hammered his engine and fireman for no particular reason. This was to be the person with whom I was required to work every day until one of us achieved promotion. Subsequent experience certainly bore out these many stories, particularly on the Bournemouth stopping trains. To driver Batten the regulator only had two positions; wide open or shut. Similarly, the brake was either off or hard on. But to counter this, never once were we late arriving at our destination.

No. 75070 being prepared for her next duty at the top of the shed. The fireman is pulling the fire through ready to build up the fire bed.

The very first morning with my new driver we were to work a train that started from Eastleigh, stopping at all stations to Southampton Central and then semi-fast calling at Brockenhurst and Bournemouth Central. It was an easy mid-morning start to prepare the engine and I signed on duty ahead of time, but there was no sign of the driver. Thinking he must be running late and eager to get the new pairing off to a good start, both the fireman's and drivers preparation tasks were completed on the Standard 5MT and the engine run outside of the shed ready to depart. Despite my concern, no one else seemed worried; it was taken for granted that the driver would arrive and the engine would go off shed at the appointed time. Equally it was taken for granted that his fireman would do the oiling and other tasks that were the responsibility of the driver.

With about two minutes to spare, there was a clatter as the driver-side cab door was unceremoniously pushed open and a leather driver's bag flung up on to the footplate preceding the appearance of a short, stocky driver with a cheerful round face wearing a broad grin and, as ever, a cigarette protruding from one corner of the mouth.

7P 5FA
34095

Above: **Another Eastleigh stalwart, No. 76066, on shed.**

Opposite: **The footplateman's office from ground level. A very clean No. 34095 *Brentor* at Eastleigh.**

No. 73065 on what appears to be a Bournemouth stopper, passing the site of Stoneham sidings with Stoneham up intermediate signal at caution. Notice also the new signal post installed ready for installation of four-aspect colour light signalling.

'Watcha nipper, get the brake off we gotta go,' was the first greeting from my new mate, and before the tender brake was fully off we were away up the shed exit road at a fair old turn of speed. Hardly a word was spoken as we took the engine into the carriage sidings to collect our six coaches and back them into the platform.

In those few minutes waiting for passengers to board, conversation across the footplate opened with an enquiry by the driver as to whether I had actually worked one of these trains, which were to become bread and butter work in my foreseeable future. Having imparted the information that I had never been past Brockenhurst before, I was told not to worry as he would keep an eye on me and show me the ropes. And, as this was my first day, he would make it as easy as possible. This last statement was accompanied by his characteristic broad grin.

The signal arm raised, giving us the right to leave, the guard's whistle summoned the drivers attention and the green flag gave indication that all was ready with the train. All of a sudden my world seemed to be turned upside down. We shot out of that familiar station like a bullet from a gun. We roared past the engine shed as we accelerated towards Mach 1 speed and very quickly I had to ply the shovel to replace some of the fire that my kindly driver, who was apparently taking things easy, had blasted out of the chimney.

Onwards we roared towards our first stop at Swaythling – or at least I thought it was our first stop. As we rapidly approached the home signal for that station, the regulator was still wide open and I gained the impression that I had misread the timetable and we were not to stop there at all. Then all of a sudden, about a quarter of a mile away from the platform, everything was reversed. The regulator was slammed shut and the vacuum brake handle pulled back to the fully on position. Railway trains, even fully braked passenger trains, do not stop very quickly and so it was that we bounded into the platform with the brakes hard on. It was in that condition that the engine and most of the first carriage passed clear through the platform and came to a stand without the facility for passengers to alight. However, it must be said that we did manage to get five out of the six carriages fully on the platform! Walking over to the driver's side, I could not help but congratulate him sarcastically on his 80 per cent success rate. He did not respond with words but just gave that broad grin and a puff on his ever present cigarette.

The guard's green flag was our cue to once again engage in ballistic activities as we set off towards St Deny's. Although the fast approach to Swaythling was just an exercise to initiate his new fireman, this general style of driving was his standard practice, but I will say that never again did we overshoot a platform.

As the name implies, this particular link worked the majority of Eastleigh's duties to Bournemouth from stopping trains calling at every station on the way: semi-fasts calling at Southampton and Brockenhurst; fast trains where we would travel to Southampton ahead of the train and relieve the crew that had worked down from London before working non-stop to Bournemouth; and goods trains to and from Eastleigh yards.

Water was always a high priority on Southern routes and trains to and from London always stopped under the water column at Southampton to replenish their tenders. Not all the work in the Bournemouth Link involved going to the seaside, as other destinations visited regularly included Portsmouth, Chichester, Salisbury, Andover and up the main line as far as Basingstoke with an assortment of trains from passenger, freight or parcels and with the whole range of locomotives to be found at Eastleigh Motive Power Depot, both ancient and modern.

Each locomotive type, or class, also had its own unique method of firing and driving. A technique that gave best results on one class would bring about total disaster on another. Within each class there were some engines that steamed better than others and some, particularly in the final years of steam operation when maintenance was being severely curtailed, that were an absolute nightmare and a constant struggle just to maintain enough steam to keep moving.

Throughout all the time I spent as driver Batten's fireman, he never once referred to me by my Christian name, it was always 'nipper' or some friendly, but uncomplimentary, name. He was indeed the hardest of drivers, but it was truly said that if you could successfully fire to 'Thrasher' Batten, then you could fire to anyone. The skills that were learnt the hard way during that period stood me in great stead for the tasks ahead as a main-line fireman.

Whatever characteristics my driver had in his driving technique, he was still the best of mates. I was *his fireman*. After a hard day, we would frequently spend a social hour quenching our thirst. He was also a keen and accomplished fisherman who virtually insisted that I join him in that sport, and I do believe he had ideas of pairing me with his daughter, who was of a similar age. Unfortunately when he introduced me to that young lady he had forgotten – or never knew – my name and had to introduce me as his 'nipper'.

Many of our duties involved visiting the other sheds at Bournemouth, Fratton or Basingstoke. All of these were much smaller than Eastleigh, and Eastleigh men were proud of their premier shed status. There would always be light-hearted banter and leg pulling between men from different sheds, and when we arrived on shed my driver would immediately go to the enginemen's mess room while I cleaned the fire. By the time I also arrived at the mess room, driver Batten was invariably in the midst of some argument that he had deliberately started, just to liven things up a bit. More than once we felt obliged to take our can of tea and retreat to our engine cab to have our meal.

Super power to shunt a parcels van at Pompey Town, also known as Portsmouth and Southsea low-level. The engine is No. 34057 *Biggin Hill.*

A panoramic view of Eastleigh station area with No. 75074 pushing stock out of the carriage sidings and back into the down platform prior to departing for Bournemouth. It is interesting to note that by the time of this picture, the route indicator clock for outgoing engines had gone and been replaced with a trackside telephone to the new Eastleigh Panel box, and both arrival and departure roads are signalled for departure. The old semaphore dummy signal was replaced with a colour light version called a dolly.

Filthy dirty GW Hall on the York–Bournemouth. Eastleigh crew would relieve Banbury men at Eastleigh.

Bournemouth Central with No. 34071 *601 Squadron* waiting with a Waterloo service.

No. 33023 on a down parcels train to Southampton Terminus at Winchester, before the platforms were extended.

Regrettably 'Thrasher' Batten did not live to retirement age, the ever-present cigarette being the prime cause of his demise. But there remain fond memories of the many antics we shared and of him demonstrating the finer points of casting with a fishing rod. These lessons took place in the cab of a 'U'-class engine at the carriage sidings of Bournemouth West, having previously worked there on a train from Southampton Terminus with the fishing rod and tackle box stowed up on the tender; he having spent an hour or so at his sport immediately before reporting for duty.

Promotion to the Holiday Relief Link meant bidding farewell to that arduous grounding in locomotive lore, but the experience was invaluable.

The original reasoning for this new link was that half the roster would be made up of regular duty turns with driver and fireman paired in the usual fashion, while for the other half, driver and fireman would be split up to cover for men on holiday leave. In practice the desperate shortage of firemen meant that, on the non-regular working, the firemen were used to cover any duty with any driver. It was in this link that such faraway places as London, Weymouth, Didcot and once all the way through to Exeter, became part of everyday working. The heavy freights to Feltham and London Nine Elms were also in the remit of these men.

Any fireman's first trip to Waterloo was an experience that stayed in the memory as a landmark. Once you had got your first fast or semi-fast trip to the Smoke under your belt, then you knew you had arrived. Henceforth you could hold your head up high and count yourself among the elite; you were a main-line engineman. To work a passenger to London and back was known to Eastleigh men as an 'up and down'; a more exact description depended on the train you worked out of London on the homeward leg. One regular turn involved the 11.30 semi-fast from Waterloo, calling at Woking, Basingstoke, Winchester and Eastleigh; hence this was termed as 'up and down 11.30'.

London had its own atmosphere unlike any other. It was, of course, busy; the closer you travelled towards the terminus, the busier it got. The outer suburbs were pure suburbia; the inner suburbs were dowdy and dirty.

No. 73110 *The Red Knight* exiting the tunnel under Southampton city centre. In the 1960s this was still very much a rundown area with a lot of waste ground resulting from bomb damage during the Second World War.

Ex-works No. 80147 on an Eastleigh to Bournemouth stopper.

Pompey Town or Portsmouth and Southsea low-level station, with Eastleigh's No. 76011 waiting to depart with a Southampton or Salisbury via Southampton working.

Charley 22 at Salisbury, August 1963. *Courtesy of Kevin Robertson*

The cramped and far from ideal conditions at Bournemouth Central Loco shed.

A 76 Standard coming on to Weymouth shed. Weymouth was the furthest west that Eastleigh men travelled.

Waterloo with a very dirty No. 34004 *Yeovil* arriving and Standard 5 No. 73119 *Elaine* waiting to depart with some very nice clean white headcode discs for a Bournemouth working.

No. 34021 *Dartmoor* leaving Waterloo with a short train, probably terminating at Basingstoke.

Woking station provided the first clue as to your destination with green London Country buses parked outside the station and green electric trains from the Portsmouth Direct route. As more and more branches swung in to join the main line, so the number of other trains coming and going from the capital increased. In addition, there was excitement either side of the tracks; Weybridge with the former Brooklands racing circuit, now occupied by bits of aircraft, on the right side. Walton-on-Thames, little Hersham and Esher with Sandown Park horse racing track right alongside the railway. Then on through Hampton Court Junction, where every driver subconsciously looked at his watch for this was the main timing point to and from Waterloo. Here a fast train would need to be about eighteen minutes away from arrival time. Then you would sweep down through Surbiton, through 'haircut tunnel', actually a main road bridge where each railway track had its own arch to go through and those arches were so narrow that if you leant out of the cab too far there was a real danger that your head would be knocked off.

Round the long right-hand curve and into Berrylands, where steam would be shut off and the brakes lightly applied to bring the speed down to 60mph. At New Maldon a 60mph speed restriction carried all the way to Waterloo. Raynes Park was considered to be the start of the run into the terminus and, like Hampton Court Junction, the point at which all main-line trains aimed to arrive at on time. From here on into the terminus traffic would be so dense that you had to slot into your allotted pathway. A late or an early arrival could wreak havoc with the signalman's job of trying to keep trains flowing in an orderly procession towards the complex that was Waterloo. Raynes Park was where the red London Transport buses made first their appearance but it was also the point where the fireman could relax his labours for most of the final stage of the journey would be coasting. This was the start of London proper where the next station, dowdy Wimbledon, bore no relation to the manicured lawn tennis courts and strawberry teas of another part of that same suburb.

Next was Earlsfield and then into Clapham cutting, where the Brighton line joined from the right and the brakes were applied again for the 40mph restriction through the curve of busy, busy Clapham Junction. By now there were lines going in every direction and commuter trains scuttling and scurrying beside you, over you and under you.

The main line from Windsor swung in from the left and at one point there were no fewer than sixteen running lines out of Clapham and each one, it seemed, had a train on it. Eight of these lines led directly to the terminus. Queens Road glided by and the tall tenement blocks and backs of factories led on to Vauxhall with smoky, dirty, Nine Elms shed down on the right and the goods yard of the same name down on the left, with the running lines we were on still on an embankment.

With reduced speed, the 'S' bend ahead of the terminus was negotiated before the whole panorama of that maze of lines and platforms called Waterloo fanned out before you. Carefully you threaded your way over the mass of pointwork and into the platform to disgorge your trainload of passengers, adding to the already teaming throng of humanity that was a permanent feature of the concourse's twenty-one platforms.

Even before the train had come to a stand, the first and most eager passengers had alighted to jog up the platform towards the exit. London is a hectic place where every second counts and such civilities as actually waiting for the train to stop before alighting are not observed, except by tourists.

Just occasionally a bowler hat city gent would raise his furled umbrella and utter a 'thank you driver' as he passed by the engine.

Having uncoupled the engine from the train, it was time to sit down and relax until a tank engine coupled on at the other end and pulled the carriages off on their way back to the carriage sidings at Clapham, where they would be cleaned ready for their next journey. The train engine was then free to reverse back to the engine shed at Nine Elms, where coal and water replenishment accompanied fire cleaning. Disposal completed, the crew could go into that dark and dirty hovel that Nine Elms men called their mess room.

Eastleigh and Nine Elms were by far the biggest Motive Power Depots on the system and each considered itself to be the premier shed. It therefore follows that the inter-shed rivalry between crews of each was particularly acute, with no holds barred in the condemnation of facilities and practices at either depot by the opposition. Entry of Eastleigh men into the heart of Nine Elms would invariably invite some derogative remark like, 'Look out lads, the country bumpkins are here,' to be countered by a statement on the state of their cabin and where overalls would get dirtier by sitting on their seats than ever they did by working an engine. However, all was largely good-natured; they were all South Western men, which is all that really mattered. Exchanges between enginemen from that other railway called the Great Western were perhaps not quite so friendly. Long after it had become British Railways, Western Region, it was still considered to be a rival and the enemy, enmity born from historical friction between the two companies.

No. 34001 *Exeter* on Nine Elms shed.

However, there were no Western men at Waterloo; they were usually encountered at Salisbury and Basingstoke, or when they worked to Eastleigh. Eastleigh men also had forays into foreign territory on through workings to Oxford, Didcot and Banbury.

Working over the Western Region lines entailed different working practices. The first noticeable difference was that the margin between the running lines was greater, a legacy of the old broad-gauge days. The signals arms were shorter and wider; being lower quadrant, they showed clear by lowering from the horizontal rather than raising.

On the Southern, quadruple track was laid such that the local and main lines in a given direction were adjacent. Thus traffic could be crossed in or out without obstructing traffic in the opposite direction. On the Western, up and down local were adjacent, as were up and down main. Thus to cross from local to main, or vice versa, meant crossing traffic in the opposite direction.

The Western Region became fully dieselised in the early 1960s, several years before the Southern and steam facilities were consequently curtailed. Working the Oxford freight as far as Didcot, the Eastleigh engine, usually an N class, would come off the train and go to the loco shed for servicing, while a Western Region diesel engine took the train forward.

Didcot shed at this time was a quiet, almost forlorn place. The doors to the shed were closed and red banners placed across each line. There were no engines and no people visible as the Southern engine made its way to the turntable. The crew turned their own engine and filled the tender with water ready for the return trip. As this was a Western shed, the meal break was usually taken on the engine. Even so, the fireman would need to go to the drivers' cabin to fill the tea can.

Entry into the shed was by a smaller door let into one of the main shed doors. Inside the shed was an array of former GWR steam locos, which had been purchased by the fledgling Great Western Society and stored inside the shed. All had received careful cleaning and it has to be admitted that the GWR engines were pretty, with their green paint, brass safety valve bonnet and copper-capped chimneys. It is just a pity that the chap who designed them had never had to work with them! Suffice to say that they were awkward to work on and the cabs were crude and spartan. They were also GW, which meant that LSW men automatically disliked them.

Fratton Loco coal stage, with No. 80139 in the centre road of the three. Coaling was done by means of a crane, whose jib is visible. This was not a grab crane. The visible 1-ton tubs would be filled straight out of a wagon and a full tub craned up and released into the tender or bunker of an engine on the adjacent road. Fire cleaning could be effected on any of the three roads, although the two roads not immediately adjacent to the crane were usually full of wagons or stopped engines. The shed was accessed directly from the coal road and the turntable was situated within the shed. This was not strictly a roundhouse as the shed structure was in fact square-shaped. *Courtesy of Kevin Robertson*

4
Night Freight
A Combination of Experiences from Many Such Journeys

The Southern Region of British Railways and its predecessors had always been a predominately passenger railway. Such freight that was generated collated at marshalling yards such as Eastleigh from the many and various local goods yards.

Because of the concentration of passenger traffic on the lines during daytime hours, it was not practical to run heavy, slow, long-distance freight trains until late night or early morning when the faster trains ceased running. From the yards at Eastleigh, several of these freight trains ran daily to places such as Exeter, Bristol, Oxford, London Nine Elms and, the most important of all, to Feltham marshalling yard. On the later working would be all traffic going to the Midlands and North, to be remarshalled at Feltham for onward travel. The Feltham freights were always heavy and quite often required a relief train to be run after the main departure.

S15 No. 30803 doing what S15s did best, coming off the up local and through Shawford with a lengthy Nine Elms freight.

The engine and crew for a lot of these trains came from Eastleigh shed and the Feltham left Eastleigh marshalling yard at about 4.30am. As the round trip was a long, slow and hard turn of duty, the crew would have their engine prepared for them; invariably an S15-class heavy freight. The driver and fireman thus only needed to sign on duty about twenty minutes before departure from the shed. It allowed time for the driver to acquaint himself with the daily alterations and traffic notices, while his fireman checked that the engine had been prepared to his satisfaction.

After a quick check of the notices to ascertain any temporary speed restrictions or other operational notices that might affect him, the fireman left his driver and walked the short distance from the office and over to the running shed, where his engine should have been stood outside under the water column, gently humming away to itself as only a steam engine with a good fire in its grate could.

The first job was a walk around the outside of the loco, checking that all was in order. Even though it was pitch black, the engineman knew where everything should be and the telltale dribble of clean oil would show that the oiling had been done. A single lit paraffin lamp over the right-hand buffer adorned the front of the engine, being the correct headcode for Feltham, and another lamp with its red filter showing was placed on the rear of the tender. On climbing the footplate, a further lit lamp would be noted hung on the tender front bracket as a spare.

The first jobs on entering the cab were to open the fire doors and check the state of the fire, and to test the water gauges so as to inspect the water level in the boiler. Ideally the fire should have been well spread across the grate, with a good spread of fresh coal over the whole fire and a few more good shovelfuls just under the fire door where the fire needed to be deepest. With the glass showing the meniscus about halfway up its length and the pressure gauge at 140psi, the fireman ought to have a good start to his night's work.

The blower was then turned on to create a forced draft upon the fire and so draw it up; and the fire iron was taken down from the top of the tender and the fire given a good rake through. Locomotive fire irons varied in length and shape depending on the job for which they were intended, but they were usually in the region of 10ft long with the shortest dart some 6 to 7ft, so on the road they were used sparingly because of their awkwardness in the confined space of the cab, especially when you had a driver taking valuable manoeuvring space on his side. It was vital on these journeys to have a good thick bed of clean fire to build upon, so having raked the fire, the fire doors were closed almost shut and the injector turned on to fill the boiler. The action of injecting water into the boiler also cooled the existing water and lowered the boiler pressure, which at that stage was not a problem and gave the fireman some leeway to build up his fire without wasting steam.

Leaving the fire to burn through a bit, the fireman next checked that he had all the tools he and his driver would require, that they had detonators and red flags for emergencies, the tender had been topped up with water and the coal pulled forward and trimmed for safety. Most important of all the tools, the fireman checked the firing shovel to see that it was in good order and was a proper Southern type and not one of those silly GWR teaspoons. If he was unhappy with any of the tools he would raid the equipment of another engine further back in the line whose crew had not yet started to do preparation work. When all was well, the fire was checked again and should have been ready to commence the building up process. With luck the pressure would be back to about 120psi and the boiler not quite full, so there would be plenty of room for manoeuvre.

Taking his firing stance of left foot across the tender fall plate, right foot towards the firebox and his body at right angles to the fire, he loaded the first shovelful of coal and by deftly swinging in one continuous movement, shot the coal exactly where he needed it on the fire. In this manner he would quickly cover the entire fire bed with fresh coal and put a further five or six under the door. All this was done in the dark, as steam engines did not have an electric light to be switched on. The only illumination was from the small paraffin gauge glass lamp, hung on its bracket, whose weak yellow light was only sufficient to light the glass tube showing the all-important level of water in the boiler.

Any other light was from the warm and soft light reflected from the open firebox, which picked out the polished brass and copper of the control handles and gauges. The boiler front, known as the backhead, was painted black and buffed up with an oily rag so that it shined. Whatever the unkempt state of the outside, the cab of a steam engine was relatively clean and kept in an orderly state.

Having put the first round of coal on, it would have been time to walk up to the front of the engine again and check that the smokebox door was tightly shut and any smokebox ash had been swept from the front area called the bogie table. Gritty ash on the front could, and certainly would, blow back into the driver's eyes once out on the road.

Regaining the cab, the fire was checked again and further coal added if required, with possibly another three or four shovelfuls put under the door. Steam pressure would be creeping up nicely and all should be well. Time to have a good sweep up of the cab floor, followed by a hosing down with the pep pipe – a device that took scalding water directly from the boiler and was used to dampen down coal in the tender or hosing down the cab. Some was also put into the bucket that all engines carried and when cooler would be used for washing hands. A lot of the fireman's work involved handling dirty or greasy items and real enginemen did not wear gloves; they only had an engine cloth constantly in their hands. Consequently, when grease-laden couplings and the like had to be handled, it was a comfort to have a bucket of hot water to wash the grease off.

If all had gone to plan there would still be a few minutes before the engine was due off shed and the fireman could take a rare breather. He closed the rear damper until it was only just open, knocked the blower back until it was just sufficient to stop smoke and flames coming back into the cab and sat himself down on the hard wooden shelf that passed as his seat, usually hinged up against the cab side out of the way when things got busy.

There in the soft, warm half-light, with his living engine softly murmuring to itself like some dozing mammoth, he could quietly contemplate the night's work to come. As he sat there surveying his little kingdom, steam rose gently from the freshly dampened floorboards, another wisp of warm steam came from the injector overflow and the smell of hot oil filled his nostrils with that certain evocative smell that only a steam engine could produce. Smoke was rising almost lazily from the chimney and in his ears would be the hum and buzz of an engine with fire in her belly quietly making steam. The fireman would also be also very aware that 90 per cent of all other males, now asleep in their beds, would give a great deal to be where he was now – before the real work had started.

His period of contemplation was not long and presently his driver would shatter the peace by clambering up the footplate steps. With a cheery, 'We all right matey,' he stowed his bag and coat in the tender locker alongside that of his fireman's.

Giving some ribald comment to the effect that it was 'all right' until his driver disturbed the equilibrium, the fireman raised himself and unscrewed the tender handbrake while his driver opened the ejector to create the vacuum that would operate the engine brake and eventually the train brake.

Having convinced himself that all was in good order, the driver checked the cylinder drain cocks were open, put the reverser into forward gear, gave a toot on the whistle just to warn any other poor soul who should happen to be about at this unearthly hour that they were about to move off, and gently eased the regulator open just a bit. The engine stirred herself, and the wheels gave a creak and a groan against the rails as if complaining about being disturbed. Steam hissed from the drain cocks at the front of the engine and as she took steam into her cylinders, the sleepy mammoth moved off gently with the merest 'chuff' from the chimney. The quarter of a mile or so from the front of the shed to the ground signal guarding the main running lines was taken at a sedate pace, and when he was satisfied that no steam that had condensed into water was left in the cylinders, the driver closed the drain cocks so the engine glided almost noiselessly up to the signal.

Beside the track, just in front of the signal, was a wooden case upon a metal post. On opening the door of this case, the fireman was confronted by what looked like a clock face, except that there was only one pointer hand and instead of numerals there were metal discs with inscriptions such as 'Up Main', 'Down Main', 'Loco Works', 'Town Yard', 'West Yard', 'East Yard', 'East Docks', 'West Docks' and several others. Eastleigh was a busy locomotive depot with engines coming and going at all hours; the train describer was an essential piece of equipment for the signalman. The pointer was turned towards the disc bearing the legend 'West Yard' and the heavy plunger beneath the clock was given a healthy thump.

Up in Eastleigh West signal box (which was geographically to the south of the station) a similar clock was located. The operation of the plunger sent an electrical signal that replicated the pointer position as set by the fireman at his end. The signalman was now aware that an engine was standing on the exit road and needed to get to the West yard.

Presently there was a muffled thud as a well-greased point blade was moved to a new position, followed by the metallic clunk of the ground signal, known as a dummy, rotating to the clear position. With a steamy sigh, the engine brake was released and regulator pushed open, causing the engine to surge forward and move smartly away over the main lines and up through No. 2 platform road. Another dummy, this time located on the lattice post of the platform starter signal, was set at clear and the engine cantered up through the deserted platform, clattered her way over the junction with the Romsey branch before swinging left into the up goods reception line and came to a stand behind a third dummy that was facing the opposite direction. They were now opposite Eastleigh East signal box, which was geographically to the north of the station and in full view of the signalman, who quickly set the road and cleared the dummy, giving the driver permission to reverse across the main lines and into West yard. As the engine, now travelling tender first, entered the yard, a shunter was on hand opposite the siding containing the train for the Feltham, swinging a white hand lamp from side to side, beckoning the driver back onto his train.

Being an experienced crew, the engine was brought to a stand right at the point of impact between the tender buffers and those of the first wagon. It was the fireman's responsibility to couple the engine to the train, which was known colloquially as 'tying on'. This was achieved by ducking under the buffers to stand between engine and wagon, and first unpinning the vacuum pipe from the engine. Thus the engine could not move as the vacuum was destroyed and the brakes were hard on. Then, taking the lower link of the heavy engine screw coupling in both hands, it was swung up and over the hook on the wagon. The link was a fairly close fit over the hook and extreme care had to be taken to ensure the coupling link was released before fingers can be trapped. Failure to observe that particular rule was painful to say the least.

Once the coupling was attached, the screw was tightened to make the coupling taut, but not overtight.

Next the free ends of the vacuum brake hoses on both wagon and tender, which were already detached from their housing, were brought together so that the bottom lugs were intertwined and the holes in the top lugs lined up for the locking pin to be inserted. Coupling complete, the red tail lamp was removed and the red filter turned clear of the light. It was not good practice to be wandering around with a light showing red for danger unless you mean it. As we already had one spare lamp lit, the newly retrieved one was banged smartly on the ground, which was a lazy, unofficial but sure way of extinguishing the flame light.

While awaiting the arrival of the guard with information about the train, the fireman again set to work with his shovel. The short journey off shed would have caused the fire to draw up nicely, and in the first mile or so of their journey he knew that he would need a heavy fire. If his fire was right, he would put on a further fifteen or so heaped shovelfuls all over it so that it could be getting hot ready for conversion into heat energy.

The guard had come on duty sometime earlier and had walked down the length of the train checking all was in good order. He noted the weight of each vehicle and the number of vehicles in the train; also any peculiarities that the driver would need to know. If there were any exceptionally heavy loads, such as steel girders, they should have been marshalled as close to the front as possible. If for some reason they were towards the rear, the driver should have been made aware that if he braked hard at the front, he was likely to be run into by the heavy weight at the back.

There were three classes of freight train: the fully fitted where all the wagons had brakes operated by the engine. These trains were in a minority and were usually the domain of fast perishable traffic. Unfitted goods were those where none of the wagons had vacuum brakes operated from the engine and were restricted to 30mph. The majority of freight traffic, including the Feltham, was partially fitted, which meant that a minimum of 10 per cent of the number of wagons must be vacuum fitted and marshalled next to the engine so that the wagon brakes could be used in conjunction with the engine, and speed restricted to a maximum of 45mph. It seems to have been an unwritten rule that if regulations stated there should be a minimum, then that is what you got and no more.

The full load for an S15 goods loco as rostered on the Feltham freights was 1,000 tons, due mainly to the long uphill drag right from the start at Eastleigh and for the ensuing 19 miles to Steventon, which is about 5 miles south of Basingstoke. This was a long, hard slog with a relatively cold engine, which would be worked flat out to get the train under way. Under railway parlance, loads were calculated in units of 10 tons – that being the original standard weight of a loaded goods wagon. Wagons weighing more than the standard were calculated and the resulting weight converted into a weight–vehicle notation. It all sounds very complex and its roots go back into history, but it was clearly understood by engine crews, who really only wanted to know how many wagons they had and how much they weighed.

Having made his calculations and taken note of anything of which his driver should be aware, the guard walked back up the length of the train. Tapping on the side of the cab, he attracted the driver's attention and informed him he had '86 equal to 99, 10 vacuum head'. The driver acknowledged this information and the guard started the long walk back. The statistics related informed the driver that he had eighty-six wagons of various sizes, whose total weight ran out at 990 tons and the brake force available was the first ten wagons. Thus there was a brake force of 136 tons of engine plus 100 tons of wagon, or 236 tons to stop the rest of the train weighing 890 tons. Heavy goods trains do not stop quickly.

The train itself would be well over a quarter of a mile long and at this point, standing in a curving siding with other trains made up on the adjacent tracks, the engine crew could not see the end of their train. In fact, they could only see the first dozen or so wagons and the brake van would be way back down the yard in sight of South Junction signal box, which quite naturally was to the east.

The engine crew were now well into their preparation. The driver opened the main ejector valve, which caused a powerful jet of steam to eject the atmospheric pressure from all the vacuum brake cylinders and thereby create the vacuum used to apply the brakes. He watched intently as the twin needles on the brake gauge rose in unison to register a maximum of 21in of vacuum before closing the main valve and opening the small valve, which maintained the vacuum at that level until it was destroyed by the operation of the brake handle. There were then two brake systems available to the driver. A steam brake operated just the engine brakes and was used when travelling light engine and for buffering up a string of loose-coupled goods wagons so that the whole train was pushing against the engine before an earnest brake application achieved by operating the vacuum brake, which acted on the engine's tender and those wagons in the train whose brakes could be operated from the engine.

Meanwhile, the fireman would have completed the build-up of the fire to his satisfaction. Each different class of locomotive had different firing techniques to achieve the best results and each fireman had his own method of working. These particular engines had a long, narrow firebox set down between the frames. The depth from the fire door to the grate was about 3ft and ideally at this stage, just prior to departure, there would be a covering of clean, well-burnt fire about 12in deep at the front of the box sloping up gently so as to be about a foot under the door. On top of this the fireman would have successively built up an overall layer of 12in of fresh coal all the way over, giving 2ft fire depth at the front and 3ft at the back. Hence the back of the fire was level with the bottom of the fire door. The fresh coal would be burning steadily with an orange, unforced flame: this would soon change.

When all was ready, the driver ensured his engine was in full forward gear and gave a short pull on the whistle to inform the yardmaster that they were ready to go. The steam pressure gauge showed 190psi, just 10 short of maximum pressure and already a white feather of excess steam was issuing from the safety valves.

Over in his cabin, the yardmaster heard the whistle and contacted the signalman at East box, who controlled the departure. East box sent a bell code to his colleague at Allbrook box, who would have the responsibility of signalling the train from the departure road out on to the main line. Allbrook acknowledged and East box set up the road from West yard to the departure road, which was parallel to the main line. There were three ground signals, or dummies, which controlled this movement and none could be seen by the driver tucked away in the siding. Having received clearance, the yardmaster walked over to a position in sight of the driver and gave a green lamp signal waved from side to side.

Releasing his engine brake, the driver acknowledged the signal with a toot on the whistle and with a cheery, 'We're off bud,' to his mate, gently eased the regulator open very slightly. With a long hiss, almost a sigh of resignation, the engine seemed to sit up as the steam, which was her lifeblood, surged through superheater elements and main steam pipes, which were her veins and arteries, before entering the cylinder and pushing the piston slowly but powerfully forward. This power was transmitted via the connecting rod to the wheels, which in turn commenced the first revolution with a groan of steel tyres on steel rails. The engine eased forward slowly, taking up the coupling slack between first and second wagon, then each successive wagon in turn as the engine advanced gently. The column of dark smoke emitting skywards from the chimney was interrupted by a long, almost lazy, whoosh as the first exhaust beat was ejected, and with a simultaneous gasp from within the firebox as the first draft of air was drawn through the fire.

As each wagon was set in motion, so the weight on the engine increased and the driver countered this by opening the regulator just a touch more until, after about 50ft or so, he could be sure that all of his unseen train was on the move. Giving the regulator a good push so that it was about a quarter of its full potential, the exhaust beat from the chimney became a pronounced chuff as the engine dug in and started to pull in earnest. Slowly the train wound its way out of the yard towards the departure road, the exit of which was guarded by two semaphore signals arranged side by side on a lattice post; the left-hand one giving clearance out onto the up relief line and the right hand to the up main. Both remained defiantly at danger.

At a steady 5 or 6mph as judged by the driver, for steam engines of this type had no speedometer, the train drew steadily forward and the signal remained on. A third of the way up the departure road and the driver was getting annoyed; the last thing he wanted was to have to stop and restart his train unnecessarily. With an exasperated, 'Come on, wake up Bobby,' he gave the whistle cord a pull in the belief that the bobby, or signalman, has forgotten them.

Far from being asleep, having accepted the Feltham goods from East box, Allbrook had sent bell codes up the line to Shawford, offering the train to him. Shawford knew that to accept would mean the goods would trundle slowly out of the yard across both down lines and had to ensure there was nothing coming on the down line or was likely to be held up by the passage of a slow goods train. Having thus satisfied his concern, he returned the bell code to Allbrook and accepted the train, while also informing his counterpart at Winchester that the Feltham was on its way.

Back on the engine, the driver had now shut off steam with some disparaging remarks concerning all signalmen and their ilk and was about to start braking his train when the right-hand signal jumped smartly upwards. 'Up main,' called his fireman. 'About bloody time as well,' responded the disgruntled driver and he gave a couple of short blasts on the whistle as a 'thank you'.

Having freewheeled for a time, the driver now had no idea whether the couplings were taut or slack, so he had to go through the procedure of opening the regulator gently at first to allow the engine to run ahead of the train, then a bit more until she was pulling again, then up to a quarter open for a while before pushing the long handle to the half-open position.

The engine dug in again and was pulling steadily; the report from the chimney was now a healthy bark as steam and smoke were forced up into the night sky. All normal verbal communication between the engine crew was by now impossible except by shouting across the cab, and all communication would now be done by recognised and well-practised hand signals.

The solid and powerful lines of Urie S15 No. 30503 at the rear of Eastleigh shed.

The whole ensemble was travelling at around 10mph as the engine passed under the starting signal and the fireman took the spare headlamp from its bracket on the tender front, walked over to the driver's side and showed the white light back along the length of the train as the engine swung out of the yard across the down main lines and thus at an angle to its train. For the first time the engine crew could see the entire length of their train, and there, a long way back, could be seen the white running light of the guard's van. At least the van was coming. Then a second white light appeared and was waved from side to side; we had a guard. The signal was answered by the fireman before he returned the lamp to its allotted place, had a look into the firebox and decided that a couple more rounds of coal would not go amiss. As steam pressure was holding well and water level was beginning to drop, he deftly put the injector on to top up the boiler. All the while he was looking back along the length of the train and as the main line swept round in a gradual right-hand curve at this point, he could observe each successive wagon as it passed under the last yard lamp and swung out of the yard to join the queue behind the engine, still with the regulator at half open.

Presently the white running light of the van swung out of the yard and across the line. The injector was shut off as the fireman turned to his expectant driver and with an outstretched hand called, 'Right away mate, all out.'

The driver, who was half-sat on his wooden seat and half-leant against the cab side, responded by pushing the regulator to its full-open position. The engine was now in full forward gear with full power and working flat out, travelling at around 10mph.

The chimney was now a source of volcanic eruption as steam, smoke and orange and red cinders were hurled up into the clear night sky. Each explosion out of the chimney was answered by a deafening 'whompf' as the blast was transmitted through the fire tubes and forced the draft up through the fire bed. The closed fire door was sucked inwards and clunked against the firebox. Everything that could rattle, did rattle. The bucket, still with its warm water, did a little dance to the vibration of the cab floor. The whole engine was shaking and vibrating under its own powerful exertion.

The springs of a steam engine were designed to withstand the hammer blow of heavy machinery; they had no concern for the comfort of engine crews, and as each successive rail joint was traversed, each wheel jolted over it. Each jolt was transmitted through the frames to the cab at the back. Each time the engine jolted in one direction, the tender lurched in the other. Cinders, tired of their skyward journey, now rained down on the cab roof and tender. Throughout all this deafening orchestration of noise and vibration – each sound was identified by the enginemen and conveyed information that his mind subconsciously acknowledged – two men were about their work, dragging their complaining train up the long and arduous climb towards their goal of Feltham yard.

After a period, the speed increased to maybe 15mph and the driver relented in his thrashing by winding the gear back to 60 per cent cut-off. This meant that steam was admitted into the cylinder for 60 per cent of its stroke. Power for the remainder of the stroke relied on the enormous expansion qualities of the superheated, pressurised steam. The blast from the chimney eased slightly, although the pyrotechnics were still in evidence and the fireman screwed his eyes against the fierce heat as he surveyed the state of his fire.

Instead of the warm orange glow, the fire would have drawn to a white inferno and the even bed, so carefully laid and built, would now have been pulled into chaos by the unrelenting blast.

Noting the worst-affected areas, he took his accustomed stance, balanced across engine and tender. Both vehicles were naturally moving in the same direction, but with very different ride characteristics that called for free and independent movement from both the fireman's legs. Loading the shovel, he threw open the fire door and in quick succession delivered about a dozen rounds of coal, all at precisely the spot where he intended. He had no time to study where to coal went, skill and experience ensured he hit exactly the right point in the fire bed without need to check and he had swung away to load the next round before the previous one had landed. Firing complete, the door was snapped shut with a metallic bang. He had already noted where the next effort was to be directed.

As always, his eye was drawn to the all-important water gauge glass and a note was made of the level just below the top of the glass.

He looked back along the line of the train and forward towards the front, just to cool his brow as much as anything else. Then he opened the fire door again, delivered ten or twelve more loads as required and snapped the door shut again. As each round of firing was completed, the smoke forced up from the chimney became dark with the unburnt gases of heating coal. Complete combustion caused light grey smoke and this, even in the dark, was a clue used to know when to put in another bout of coal.

Three or four heavy bouts of firing brought the fire back to order. The gauge glass was now down to three-quarters of the glass, steam was holding at a steady 185psi and it was time to start managing the boiler water level. With a well-practised hand, the water cock was knocked open, and almost in the same movement the corresponding steam valve was open and with a clunk and a whirl the injector sung away; the telltale sign that it had picked and was working correctly.

As it was only required to replace water lost in continuous evaporation, the injector was adjusted to deliver a minimum flow of cooler water into the boiler. This was another excuse for the hot fireman to take a breath of fresh, cool air. Although this trip was on a clear autumn night, in the depth of a cold winter, the fireman would be half roasted at the front where he faced the fire and frozen at the back because they were not moving fast enough to prevent an icy cold winter wind come over the tender and into the cab.

While this struggle between man and machine was taking place, the driver was stood looking at the road ahead, apparently unconcerned by the exertions of his fireman. He was well aware of the effort involved from the days when he too was a fireman. He would instinctively know what was happening on the other side of the dark cab. As the fire doors were opened, he might just cast an unseen glance at the pressure gauge to reassure himself that all was well, otherwise he had total confidence in the abilities of *his* fireman, just as the fireman relied on the skill of *his* driver. So with one hand holding the cab-side cut-out, steadying himself against the motion of the engine, and the other hand, with engine cloth neatly folded into the palm, resting on the regulator, he was watching the way forward and listening to what his engine was saying. The rhythm from the chimney was a good indicator of speed and power output.

As speed increased to around 25–30mph, the cut-off was reset at 50 per cent and the locking lever put in. The engine would not be eased any more, and 30mph was as fast as they were likely to go until the long drag was beaten. The fire storm from the chimney ceased with just the occasional orange cinder given free passage out into the wider world.

With the maelstrom over, firing could settle down into the more usual method of 'little and often'; no more than six shovel loads at a time, but often enough to counter the consumption of the hard-pulling engine, and at least the fire was likely to stay within the firebox and not straight out of the chimney!

Steadily onwards they forged their passage, through a dark and deserted Winchester station, the bark from the chimney rattling the windows and echoing around the empty station buildings. There was an exchange of waves with the signalman, who had come to his window to see the Feltham safely through his domain.

They clattered north past Winchester Junction, where the line to Alresford and Alton diverged to the right and into Wallers Ash tunnel. Although this was a fairly short tunnel, the heavy exhaust beat and relatively slow speed caused steam and smoke to rebound from the tunnel roof and into the cab. This was an unpleasant experience, which had both crew members leaning over the side of the cab for fresh air as soon as they emerged at the far end. Up past the goods loop at Wallers Ash, where previously many a frustrating hour had passed on a slower-moving freight that had been sidelined to allow faster traffic to overtake. Tonight there was nothing chasing their tail and they had a clear run through.

On through Micheldever they trundled, past the goods yard cut into the chalk hillside; a ridiculously large facility for such a small community. During the dark days of the Second World War, storage tanks of aviation fuel were set deep into the hill and the sidings served the delivery of that vital commodity.

Now much of the yard was taken over by commercial use. Several long sidings adjacent to the running lines were full of condemned coaching stock awaiting their turn to be taken away and scrapped.

With a pull on the whistle, the labouring engine stormed into the first of the twin tunnels at Popham, just north of the station, then onwards through the deep cutting at Roundwood with its little signal box dwarfed by the dark and wooded cutting sides. It was an eerie little box guarding the entrance to that black and menacing aperture, which seemed like the haunt of evil things, but was in fact Roundwood tunnel. Another shrill whistle and they entered the black hole regardless of any evil spirits they may run over.

Out the other end and the long drag was nearing its conclusion. In the moonlight, the parallel rails ahead could be seen to level out just past Steventon box. Finally, the engine pulled its train on to the level track that led to Worting Junction. As the train came over the bank, speed increased and the cut-off was eased to 45 per cent, then 40. Speed was rising to around 40mph and the regulator was half shut.

Worting was the point where the busy main lines from Bournemouth met their counterparts from the West Country and joined together as a four-track route all the way to the capital. It had long since become a fast, flying junction where the up line from Bournemouth was taken up on to on a girder bridge to cross the Salisbury lines before diving down again to join the busy London route. For a passenger train, the flyover was a great bonus as the speed over the junction could be maintained at 60mph; for a loose-coupled freight, it presented a complication.

Running freely along the level embankment, the engine crew waited to feel the gentle tug from the rear of the train as the guard applied his handbrake. With the long train running up the gradient to the flyover and down the other side, it was essential that all the wagons were running as a combined unit and all at the same speed. If the train was left to its own devices, the front part would slow down as it ascended the gradient and the rear part would effectively crash into the front, possibly causing buffer lock and derailment. If the train did get up the rise, the front part would increase speed as it descended the other side. With the rear slowing and the front accelerating, there must be a point where the wagons were running with slack couplings. The quick increase in speed at the front would snatch at the coupling and either pull a drawbar out or snap the coupling and divide the train. Either scenario was frowned upon by the 'authorities' as it meant the busy main line would have to be closed for many hours while the resultant mess was sorted out.

Hence the usual method of safe operation was for the guard to apply his brake hard and thereby cause all the couplings to become tight and cause the train to all be moving in unison. The driver could then keep steam on and drag the train both uphill and down the other side, and any increase or decrease in speed was uniform throughout the entire train.

On this particular night, the telltale tug never happened. The guard in his cosy warm van, jogging along way back from the scene of the night's action, seemed to be otherwise engaged. The flyover was getting closer and the driver had to make a decision to revert to plan 'B'. If he could not have all the couplings taut, he must ensure the opposite state and get all the buffers together and pushing against each other.

Shutting the regulator entirely, he flipped off the retaining catch and partially applied the engine steam brake, just enough to retard the engine and cause the whole train to gently buffer up together. Using his experience to judge when the whole train was running buffer to buffer, he applied his brake with greater force so his engine was pushing back against the great weight of the fickle train that was now trying to outrun the engine.

While his driver demonstrated his skill in controlling a heavy freight in adverse situations, the fireman could at last relax his exertion for a while. He took the opportunity to put the injector on full to replenish the boiler ready for the jog up the easier part of the route. Also, he had a good fire and as the driver was not expected to use the same quantity of steam on the easier section of the run, he should be able to relax just a little bit. Injecting cooler water into the boiler had the effect of lowering boiler pressure and not wasting hard-earned steam.

The engine led the way up the incline and with a hollow rumble, passed over the steel girder bridge as the main lines passed underneath and swept away, bound for the west. Off the bridge and the engine dipped her nose downwards as the rails descended the other side. With ever more of the train coming on to the descent, it pushed ever harder against the engine, causing the driver to react by reapplying his brake more and more until it was full on.

Sweeping down the incline and past the signal box, a stream of sparks poured from the brake blocks, while the acrid smell of metal rubbing against metal infiltrated the cab. On to the level track again and the brake could be eased gradually while both men strained ahead to get early sight of the signals controlling Basingstoke, just 4 miles ahead.

From this point most of the semaphore signals were on gantries directly over the line to which they applied. On a clear run there should be two little green lights above one another, but tonight the top light was green as expected, while the lower one was yellow; Basingstoke had kept its distant signal on. Something was happening ahead and they had to proceed under caution, ready to stop if the home signal was at danger.

As might be expected, the driver cursed whatever was ahead, all signalmen that they should dare impede his progress and especially all guards that they should even have to exist. He reapplied the engine brake again to be sure the whole train was still all buffered up, unlatched the train vacuum brake and brought the brake handle down partially. With a rush of air, the vacuum in the train pipe decreased and one of the twin needles of the gauge dropped from 21 to 18 as the brakes on the tender and then the front wagons began to embrace the wheels. After a short while the brake pressure was increased and the gauge dropped to 10; the brake was gripping the train in earnest. Four miles is enough room to stop comfortably but there was not a lot to spare, so it was best to get control of the train quickly. All the while, the offending signal was getting closer as speed steadily decreased. Once they were past the distant signal, they would have to assume they would be required to stop.

Earlier, in Basingstoke signal box, the signalman had been faced with a decision. He had a freight coming from Reading, which needed to cross the main lines and continue on its southerly journey. Having the Reading freight there on his doorstep and with nothing heard of the Feltham, it seemed an obvious decision to let the Reading go. But as a particular law dictates, almost immediately the Western Region train started to pull across the junction, the bell code rang from Worting offering the Feltham. He had no choice but to hold the main-line signals at caution and slow down the London-bound train. Intently he watched the track circuit light on his track diagram, willing it to go out as the Reading cleared the junction. After what seemed like an hour, the little red light extinguished and the signalman sprang into action, resting the route and pulling the signal levers back in rapid succession. First the advanced starter, then the starter, home and outer home. When all were cleared he could pull the distant to give the driver notice of a clear run through, provided he was in time and the driver had not already passed it.

Meanwhile, back on the engine the fire doors had been opened wide and both dampers shut to keep steam pressure below the maximum of 200psi. The route took a long shallow curve to the right before Basingstoke station and this obscured the driver's view ahead. With left-hand drive he could only see down the length of the boiler and the road immediately in front; he thus relied on his fireman to relay signal information. The rule book clearly stated that the driver himself must cross the footplate and personally observe all signals, but the rule book was written for people who needed to be told, and after all this was *his fireman*, who could read the road as well as the driver. With the glare and heat from the open fire door lighting the cab and the squeal of metal brake blocks on the complaining wagons behind, the fireman leant out of the cab to survey the scene. Up ahead was a column of steam and smoke indicating a train pulling hard out of Basingstoke towards them. Guessing that this was the probable cause of their delay, he searched out the two signals against the backdrop of the street lamps and other illuminations that emanated from the townscape.

Predictably, the two small lights were set at red over a yellow. Half turning to his driver, the fireman held his arm straight outstretched, palm down and shouted, 'She's on.' The driver acknowledged and dragged the brake handle down further. Studying the approaching signal that was still some way away, the results of the signalman's effort became apparent. The red light changed to green. Immediately the fireman turned to his mate and gave the clear by holding his outstretched arm up to indicate, 'Right oh mate, it's off,' then holding one finger up, he gave a qualifying 'one', meaning that the home or stop signal only was off and the distant still on.

The brake handle was pushed up to the off position, just as his fireman observed the distant signal also clearing and the yellow light now changing to green. Turning in towards his mate again, he raised his arm up with two fingers extended. The call of, 'Right away mate, two,' would allow the driver to open the main ejector valve to blow off the brakes. Watching his gauge waiting for the train pipe needle to come back to its fully off position, he gently opened the regulator to run the engine ahead of its train again. As the weight was successively taken up, the regulator was opened more and more while the fire doors were shut and the rear damper pulled open.

Presently the characteristic panting of a GWR engine was heard approaching. As the two trains passed in the night, there was no exchange of greeting between the Southern and Western engine crews. On the slightly falling gradient, speed

A visitor from the Midland, 9F No. 92134 at Eastleigh.

was soon restored and a good 40mph was reached before they passed by the quiet engine shed that was but a fraction the size of Eastleigh and up through platform 1 of the station, over the junction to the Reading and the Western, then jogged on up that boring section of main line towards Hook.

The road was easier now and the fireman could relax a little more as the engine was not working so hard, and they cantered along at the maximum speed of 45mph for such a train. Through Winchfield and Fleet, which was the racing section where fast passenger trains regularly travelled at the maximum permitted line speed of 90mph. If the train was running a little late, a driver would try to make up time to take up the allotted pathway in the busy approaches to Waterloo. It was not uncommon for drivers to 'mis-read' the speedometer fitted to modern engines and push through Fleet at close to 100mph.

Such heady things had no place in the workings of heavy freight trains and the Feltham kept steadily to around 45mph as judged by its driver, on through Farnborough to Brookwood with its massive London overspill cemetery and so down the long straight to Woking. Both enginemen hung on tightly as the engine lurched over the rough riding junction with the joining Portsmouth line. The engine rolled one way while its tender lurched the other, and even at slower speeds it was no time to be stood up trying to fire.

Just past Woking, steam was shut off and the train freewheeled towards Byfleet before the brake was applied to reduce speed ready for the left-hand turn at West Byfleet that would take the train off the main line and the sharp drop down towards Addlestone and Chertsey. Being off the main line, their speed did not rise much above 30mph as they threaded their way towards the next junction with the Reading line at Virginia Water.

Here the station was on a tight curve from the Chertsey direction and the train was brought to a stand at adverse signals in the platform. Clearly they were waiting for something to pass on the more major line to Reading. Meanwhile, as soon as the train came to a stand, the driver jumped off his engine with a, 'Just goin' for a Jimmy mate,' and disappeared in the direction of a sign proclaiming 'Gentlemen'.

Presently a diesel-hauled parcels train purred through the station without stopping. Once clear the signalman reset the road and with a clunk the platform starter signal raised its arm to let the freight go on its way. Having no driver, the fireman gave a soft toot on the whistle and was amused to see his colleague running back to the engine still adjusting his attire.

From a dead stand, the start from Virginia Water, on the tight curve, was very difficult with a long and heavy train. Gently and with full gear, the driver opened the regulator. The engine moved forward a pace or two, then stopped. The regulator was opened more, but still there was no forward movement and as the pressure of steam increased in the steam chest the wheels lost adhesion on the rail. With a violent roar from the chimney sending steam, smoke and cinders rocketing up into the night, the driving wheels slipped, causing the whole engine to shake and shudder before the driver could shut off steam and regain control.

Putting his engine into reverse, the driver gently backed into the train for a short way so that a lot of the wagons were buffered up. This would afford a vital few yards when the couplings were stretched again as the engine moved forward. However, this was a tricky operation as too much of a snatch could again break a coupling. Resetting his engine in full forward gear and opening the sanders to assist adhesion with grit on the rail, he again applied steam. The engine moved forward as it, not over gently, took up coupling slack. With loud and long blast from the chimney, they slowly edged forward with the wagon wheels screaming against the curvature of the track.

With a consoling, 'Come on girl,' the cavalcade crept over the junction with the driver watching his steam chest pressure gauge and listening to the engine. As soon as she started to slip, steam was shut off to counter the effect and then quickly reapplied before the train could drag to a stop. By pumping his regulator in this fashion, the train inched over the junction and began to pick up speed again as the wagons were successively drawn clear.

Over the last few miles, the fireman took the fire irons and raked the fire to ensure it was all burnt through ready for cleaning. Returning the iron to its allotted place on top of the tender needed be done with care. As the iron was withdrawn from the fire it was glowing red hot. It was also about 10ft long and not easy to manoeuvre in a rocking engine cab. He would be most unwise to allow contact with his driver.

At length the train trundled over the level crossing, through Feltham station and into the arrival road at Feltham yard. Almost immediately a shunter appeared to uncouple the engine. A ground signal cleared and the engine clattered down the engine release road and into Feltham engine shed. Dropping down over the fire cleaning pit and stopping opposite a water column, the Eastleigh crew were relieved by a young pair of Feltham men, who would clean and service the engine while they walked across to the enginemen's cabin for a wash, can of tea and food.

Three and a half hours' work had demanded the highest standard of skill by the driver, while his fireman has used some 4 tons of coal and about 4,000 gallons of water. After they had a half-hour food break all they had to do was the same thing in the opposite direction.

5
Boat Train Experience

A filthy dirty No. 34034 heading into the Eastern Docks (Old Dock) prior to working an ocean liner service. As she is not carrying a headboard, this is probably a second or third working from whatever liner has arrived. The first such working would have carried the first-class passengers up to the capital and usually had a headboard proclaiming the shipping line that operated the liner.

In the early hours of a summer morning the Cunard liner *Queen Mary*, using the high tide, berths at the Ocean Terminal in Southampton docks.

Aboard are the usual transatlantic voyagers, including many American tourists intent on the grand European tour, which starts at London.

A leisurely breakfast will be consumed by all before first-class travellers join the first of several boat trains taking disembarking passengers on the final leg of their journey to the capital. The initial departure, being a named train, is deemed to be the most important and carries additional attendants to cater for the needs of first-class passengers and the locomotive invariably carries a headboard.

While the shipping line panders to the needs of its departing customers, the railway readies itself for a fast run up to Waterloo. A dock shunting engine delivers the train of sparkling clean, empty coaches into the train shed adjacent to the waiting Cunarder, before guards and attendants prepare to receive the influx of curious passengers.

At Eastleigh, the rostered engine crew sign on duty to prepare their engine, a task for which they are allowed one hour before taking the engine, tender first, down the main line and into the docks. Having made the usual perusal of the working diagram, the fireman leaves his driver to his more in-depth study of the timetable and walks across to the shed to find his allocated machine.

Having spent the last couple of days 'on shed' with the fire out, undergoing boiler cleaning and other routine maintenance,

Southampton Old Docks. No. 73113 *Lyonesse* is waiting to leave with a boat train on Special Duty number 261.

the engine is berthed at the top of No. 5 road. In the gloom of the shed it is apparent from first sight that the engine is cold and the fire has only recently been relit.

As usual at this period, the engine is in a filthy dirty state with the green, lined paintwork completely hidden under a thick greasy black layer of honest grime. Someone has wiped an oily rag over the number on the cab sides to reveal the cream-coloured numerals 34098, similarly the red scroll of the nameplate fixed on top of the running plate has been wiped over so that the raised brass letters of the name *Templecombe* are visible.

Neither boilersmiths nor firelighters are renowned for their tidiness and the footplate bears witness to that fact. Oil-laden cotton waste mingles with small coal and coal dust that has been discarded on to the cab floor. The whole interior is dusty and dirty, with the boiler pressure barely showing 80lb of steam. In the space of one hour this cold, unkempt mass of heavy machinery will have to be transformed into a living vibrant machine with a cab fit for proud men to work in.

Having stowed his coat and bag in the tender locker, the fireman's first action is to open both dampers and push the heaped fire over as much of the grate as possible before adding more, good lumps of coal. Leaving the fire to draw and burn through, the two oil bottles, each containing 5 gallons, together with the smaller paraffin bottle, are taken over to the stores and filled. The full cylinder oil bottle is put up on to the running plate ready for the driver to use and the bottle containing the thinner engine oil is taken back into the cab, where the oil feeder is filled and placed on the shelf over the fire door. From this warm position, the full feeder will itself become warm and warm oil flows better than cold.

Another check on the fire and the addition of coal as necessary precedes the check on tools and equipment. In their tunnel on the fireman's side of the tender should be the fire irons consisting of dart, pricker and clinker shovel. The coal pick should be in its bracket on the tender front and the ashpan rocker bar is clipped into its lugs next to the fire doors. There should also be a rubberised bucket and hand brush, four paraffin headlamps and four white headcode discs, colloquially known as 'boards'. In the tool locker will be an assortment of spanners, including the all-important one for the gauge glass nuts; two spare gauge glasses with their rubber sealing rings; a spare gauge glass lamp; a flare lamp and, most important of all, the metal canister containing twelve current issue detonators and two red flags for emergency use. The fireman will also take particular note of the firing shovel to make sure it is in good order and of a pattern he prefers to use.

Having noted any shortcomings in the tool list, the lamps are next on the list of jobs. Each headlamp wick is trimmed and its paraffin container filled, as is the gauge glass lamp. Although this particular class of engine is equipped with a steam-driven generator for electric lights, spare paraffin lamps are carried in the event of a breakdown. The flare lamp is a crude piece of equipment from the very earliest days of steam locomotion. It consists of a strong metal container with a spout, not unlike a miniature watering can to look at. In the spout is a thick cotton wick, the end of which is immersed in the paraffin held in the container. The lit wick gives a good flame, which is used to highlight those dark places underneath the engine. Fire, be it in the firebox or in the form of paraffin lamps, is the only means of illumination on most engines.

By now the driver has arrived at his engine, all togged up in dirty overalls and oil-stained hat ready to take his part in the engine preparation.

While the driver is up on the running plate filling the Wakefield cylinder lubricators, the fireman is again attending his fire, progressively building it up as the steam pressure rises. Cylinder lubricators are followed by general lubrication at ground level, which the driver achieves using the warmed oil feeder; then the dirty part of the exercise. On a three- or four-cylinder engine the inside motion has to be oiled from under the engine, and to gain access to all the oiling points the engine has to be stopped so that the inside big end is in the correct position. Thus, with the driver watching the position of the motion, his fireman carefully eases the engine forward until, with a yell of 'whow' from the driver, the engine is set right.

Refilling his feeder, the driver disappears under the engine to tackle that oily mass known as the inside motion, leaving the fireman to walk over to the sand furnace and collect a heavy chute full of hot sand, which is emptied into the engine sand boxes situated up on the running plate.

With the driver engaged under the engine, the fireman busies himself with checking all around the engine to ensure all is in running order, opens the smokebox door to check that the char has been cleaned and the door sealed properly before sweeping any residue char from the front framing. Coal in the tender needs to be trimmed to make sure none will spill off when the engine is on the move, the tail lamp placed on the bracket at the front of the engine and the headcode denoting a destination of Southampton docks put up on the tender brackets. All the while a periodic check is kept on the state of the fire and fresh coal applied to give a good depth of live fire.

At about this time two shed labourers arrive with a barrow, upon which is balanced a large metal plate. This is hauled up on to the front of the engine and fixed on the lamp brackets across the smokebox door. The headboard proclaims that this is the 'Cunarder'.

The new ocean terminal in Southampton Old Docks with No. 34095 *Brentor* on the United States Line boat train entitled 'The Statesman'. It is mostly first class, including a green-painted ex-Pullman car.

When all oiling and other preparation are complete, the engine is moved out of the gloom of the shed and stopped by the water column to fill the tender and make the final checks of the brakes, both injectors, sanders, generator and other equipment.

Having satisfied himself that all is well, the driver returns to the enginemen's cabin to wash and change out of his 'dirties' and into clean overalls while his mate empties the ashpan and commences cleaning up the cab. All the loose coal on the floor is either swept over the side of the cab or shovelled back into the tender. The floor is then swept clean and the whole cab hosed down with hot water from the pep pipe. The boiler backplate is then wiped with oil to make it shiny black and the brass and copperwork buffed up to contrast with the black background. Windows are cleaned with newspaper and with the engine now humming away to itself as a content living machine with a good fire in her grate, 200lb of steam pressure and three-quarters of a glass of water, all is ready.

One last job remains, when the cleaned up driver returns; a last trip to the stores to hand in dirty and oil-soaked engine cloths to be replaced with clean ones.

Exactly one hour after signing on duty, engine and crew are waiting at the shed exit ready for the ten-minute run down to Southampton, during which time the fireman will continue to build his fire with fresh coal. The fire in the wide firebox will be built in horseshoe fashion, deep and heavy at the back under the door and especially around the back corners, tapering down along the sides to a lighter, brighter fire at the front and centre.

Entry into the dock complex is a cautious journey across several ungated road crossings, each guarded by a railwayman with a red flag stopping the road traffic, until finally the engine enters the Ocean Terminal train shed.

Alongside the platform waits the eleven-coach train consisting of nine carriages and two large luggage vans. A moderate load for the engine. On the platform there is a bustle of activity as mountains of luggage are stowed into the two vans while intending passengers, mostly American with an array of cameras round their necks, watch proceedings and take in their first glimpse of 'dear old England'.

As the engine reverses into the platform and on to its train, a large portion of those passengers, mostly the males, crowd forward in almost disbelief. No throbbing diesel of massive American proportions to whisk them on their way, but a rather dirty steam engine and one manually fired with coal at that.

Cameras are raised to eye level and yards of film expended on this unexpected sight, while one or two of the more intrepid souls look through the open cab door as the crew make their final preparations. Amid the buzz of American accents, one concerned, well-fed and well camera-ed traveller asks the driver earnestly if they are actually going to arrive in 'Lun Don' that day, while a high-pitched female American voice addresses her partner, 'Hey Honey, ain't that just sow quaint.'

Allbrook with the entry and exit for Eastleigh East yard obscured by No. 34009 *Lyme Regis* on an up Bournemouth to Waterloo working. At Allbrook the railway leaves behind the urban scrawl of Southampton and Eastleigh and embraces the countryside of rural Hampshire.

A down vans train signalled onto the local line at Shawford behind No. 34001 *Exeter*. Sadly, the goods yard has been completely lifted by this date.

Eventually the luggage mound disappears and the last Stetson hat and fur coat is encouraged to take their seat, all the while the driver impatiently studying his watch. Five minutes late already and they haven't left yet!

Having satisfied himself that all are on board and all doors closed, the rear guard blows his whistle and gives a green flag to the front guard, who responds with a whistle and, turning to the driver, holds out his green flag to show all is clear. In keeping with this tuneful interlude, the engine whistle is given a long pull to warn the crossing attendants to clear the road crossings.

Slowly the engine pulls forward, taking its train on the first tentative half mile towards its destination. Every opening carriage window seems to have a foreign head poking out and on this initial stage it appears to them that they were right in thinking they were in for a slow ride.

Ahead of the train, the massive wooden gates guarding the entry into the docks are swung open and two men equipped with red flags and hand bells stop the traffic on the busy road that forms the interface between dockland and the railway network.

Slowly the near 500-ton train and engine wheels across Canute Road, to the annoyance of road users that have had their journey interrupted, and on to the railway proper. With the engine across the roadway, more power is applied and the exhaust from the wide chimney becomes audible as the train makes its way past the Terminus station and on towards the junction with the Bournemouth lines at Northam.

Running under clear signals indicating a pathway directly on to the up main line, the junction is traversed at around 30mph. With the junction cleared, the driver calls across to his fireman to see how the engine is steaming and receives a 'thumbs up' sign, with his attention directed to the boiler pressure gauge with its needle just under the red line indicating maximum pressure.

A check on his watch gives rise to the driver's comment that, 'We're goin' to 'ave to run 'em,' as he braces his foot on the footrest and heaves on the stiff regulator, pulling it halfway open. With steam being applied through the fully open first regulator valve and the gear in 50 per cent cut-off, the engine digs in for the long uphill pull. Accelerating steadily past Mount Pleasant Crossing, the driver again braces himself and gives another heave on the regulator so that it is in the fully open position, before reducing the cut-off to 40 per cent.

On the other side of the footplate, the fireman has commenced his preferred style of firing, little and often. No more than four or five shovels full at a time, and noting where the next rounds will be required. By now the firebox is a searing white hot inferno with the fire banked level with the top of the 'half door', which is a hinged flap raised to block the lower quarter of the fire hole and reduced secondary air flow. The fire is now at its maximum depth and will be constantly fed every couple of minutes where brighter spots indicate a thinning of the fire; along the sides; under the door and always constantly built up in the back corners. This later exercise requires the loaded shovel to be turned through 90 degrees inside the firebox, so that the fireman's hands are very close to the fire. With the half door in place, the area within which the fireman has to direct the fresh fuel is greatly reduced, requiring a degree of skill and dexterity to place coal in the exact point that is required. Eastleigh firemen do not wear gloves.

With the engine pulling hard and accelerating against the gradient, the train roars through St Denys towards Swaythling, where a townscape of light industry and the backs of meaner housing gives way to a more residential area. The platform at Swaythling is built on the level and is the only very brief respite against the continuous up grade for the next 19 miles. An extra 2 or 3mph are gained through the platform.

With speed gradually rising, a pull on the whistle as they pass Eastleigh engine shed – not particular necessary, just a greeting to any of their colleagues who happen to be about – before rounding the curve to Eastleigh station. Another long pull on the whistle and they charge up the main through line at around 55mph with the gear locked at 35 per cent.

No. 34013 *Okehampton* with a Bournemouth service passes the closed and lifted goods yard at Shawford.

All the while the driver is either stood; leant or half seated, according to his custom, with left arm against the open cab side window and right arm resting on either the brake handle or the regulator. He appears motionless as he gazes ahead constantly, reading the signals, observing the line ahead and the motion of his engine. His eyes are continuously checking his gauges and his ears are attuned to the sounds of the engine, listening for anything unusual in that deafening cacophony of noise that prohibits normal oral communication across the footplate.

Conversely, his companion on the opposite side of the cab has no time to be still. His work is of unceasing activity. When he is not feeding his hungry furnace or adjusting controls, he is also looking ahead to sight those signals that his driver will not see in good time. This is an often and welcome interlude as it gives an excuse to lean out of the cab window for a brief, cool, refreshing draft against an over hot face. He is also listening to his engine. The steady hum of a singing injector denotes that it is working properly and water is being delivered into the boiler, keeping the water level up against evaporation as vast quantities of steam provide the raw power to the three cylinders. An increased roar from the firebox probably denotes that an area of the fire is getting thin and needs his urgent attention.

And always the noise and the heat. The constant drone of metal wheels on steel rails. The clickety-click as each wheel passes over the rail joints. The vibration of heavy machinery moving quickly and the rolling and pitching as the engine negotiates various junctions and the tender always seeming to be snaking along behind with riding characteristics totally different to that of the engine.

Every so often a bell rings out in the cab from the Automatic Warning System telling the crew that the approaching distant signal is clear. Not especially necessary in fine weather, but an asset in fog, for which London was renowned. If the distant signal is at caution, a buzzer replaces the bell and this has to be acknowledged by the driver cancelling the indicator, usually with a few selective words of enquiry as to why he has not got a clear road!

Once past Eastleigh, the urban sprawl gives way to the lush Hampshire countryside as the line follows the valley of the River Itchen. The speed rises to around 60mph by Shawford and will stay at that rate for the rest of the climb towards Basingstoke. Into the deep cutting with its several tall bridges, carrying roads over railway; through Winchester and beyond, then out of the cutting and on to a long embankment taking the line across the Itchen Valley; past Winchester Junction and into Wallers Ash tunnel.

The 'big engine side' of the coal stage with two engines off different boat trains worked to Southampton Docks from Waterloo. The front one off the Cunarder is filthy dirty No. 34004 *Yeovil* having her fire cleaned, followed by an unrebuilt Bulleid off a Union Castle line boat train. The piles of ash and clinker from the many engines undergoing disposal were constantly cleared by shed labourers employed for that task. The gentleman in the picture is a driver in the shed engineman gang. There were two pairs of men on shed engineman duties round the clock before the rundown of steam and its facilities. They worked regular shifts of 2–10pm, 10pm–6am and 6am–2pm. In the shed engine gang I was paired with driver Jimmy Thomas and the gentleman in the picture along with his fireman were our opposite pairing. Also in the corner of the office block, above the second engine and to the left of the plume of steam, can be seen the semaphore signal that was in fact facing towards the MO's office in the Carriage Works and whose sole function was for eyesight testing. It was operated on demand by one of the clerks in the first-floor office.

At the north end of the tunnel the lines are quadrupled to give passing loops for faster traffic to overtake slower trains. In the up loop, awaiting the passage of the boat train, is a parcels train bound for Basingstoke with another Eastleigh crew aboard. The fast-moving boat train gives a series of whistles as it thunders up beside the stationary parcels train, which replies with an equally tuneful greeting and mock salutations as the engines pass.

The country becomes distinctly more wooded as they pound their way towards Micheldever station, which is actually nowhere near Micheldever village, climbing steadily through the twin tunnels of Popham and Mitcham; into the deep, wooded cuttings and the tunnel at Litchfield before finally climbing out of the Hampshire Basin and on to the flatter terrain at Steventon.

The speed increases quickly as the train pulls over the top of that long drag from sea level at Southampton, prompting the driver to decrease the cut-off to 25 per cent before easing the regulator. Steam is shut off and brakes applied gently to bring the speed down to the permitted 60mph as the train rattles over Battledown flyover and down to the junction at Worting, where the train swings out on to the up fast line. Through the junction and full power is applied again, with the train accelerating away on a long sweeping right-hand curve towards Basingstoke station, 3 miles distant. Having reached 75mph, the engine is eased back to maintain that maximum line speed through the complex of lines at Basingstoke, but is immediately opened up again as soon as the engine is over the junction with the Reading line.

The long straight run to Hook sees speed up to into the high 80s with the gear back to 15 per cent, as the gradient is now gently downward. The maximum line speed of 90mph is attained before Winchfield, but today time needs to be gained, so a liberal interpretation of the speedometer reading is called for. On the sharper dip down through Fleet, the engine is let fly and 100mph is nudged before being eased back and speed allowed to fall slightly on the level track to Farnborough.

Instinctively both enginemen move away from the cab windows as, with a 'whomp' of pressurised air and a rattling of those windows, the engine tears through the narrow 'haircut tunnel' to the north of Farnborough station and onwards towards the quieter, wooded, reverent country that is Brookwood. On the east side of the railway lays the huge cemetery where the mortal remains of many Londoners lie at peace.

By comparison there is no such peace or quiet on the engine as she powers around the long curve to reveal busy Woking in the distance. Still travelling at the mid to high 80s and at about a mile from the station, a long whistle gives warning to station staff and waiting passengers that this train is travelling fast. This is vital as every track passes through a platform and every platform is always well populated with intending passengers for London-bound trains. Joining the main line from the east is the Portsmouth Direct route and the junction of the two is notoriously rough riding.

A quarter of a mile away and the fireman hangs onto the whistle cord continuously. With a wild pitching, rolling action, the engine rides the junction and with steam on, whistle screaming, and adrenalin surging through both enginemen, they thunder through the platform. Waiting passengers have sensibly taken several steps away from the platform edge, but cannot help looking in awe at the flying machine that has shattered the calm, while at the same time men hold their hats on and ladies hold down their skirts. An explosion of noise, a rush of displaced air and then they are gone; London bound. Over the now undulating track, eating up the miles through Byfleet; West Byfleet; Weybridge; Walton-on-Thames; Hersham. The regular clip, clip of the firemen plying his shovel becomes less regular as he can afford to slacken his efforts.

Esher leads to Hampton Court Junction, where the driver, who has periodically consulted his timepiece, now makes a calculated study. He should now be eighteen minutes from his booked arrival time at Waterloo and it is important to be on time as the train enters the busy flow of traffic commencing at Raynes Park. He signals to his mate that they are still one minute down despite their effort to regain all the lost time, and he allows the engine to run unchecked through Surbiton and Berrylands. The commencement of the 60mph speed limit from New Malden to Clapham means a gentle braking to bring speed down to about 65 with the hope that the signalmen on the route will slot them into their intended pathway.

Raynes Park and the signals are all pulled off clear. The start of the approach to London and the end of the fireman's constant labour. He can now afford to lean out of the cab window and feel the cool air about his face as he takes in the 'London' atmosphere. Wimbledon, and still the road is clear; on towards Earlsfield and then heavy braking to bring the train down to 40mph around the tight curves of Clapham Junction station. Once clear of that station, the engine is opened up again to accelerate through that maze of tracks, hither and

A USA-class tank doing the job they were bought to do: shunting in Southampton Eastern (Old) Docks before they were displaced by the Ruston & Hornsby diesel shunters.

thither to Queens Road. Vauxhall, and the signals are showing caution. Heavy braking brings the train almost to a stand in front of a gantry of colour light signals that is Waterloo inner home. Before the train has quite stopped, the red signal changes to yellow with an indicator registering the legend '12', allowing the driver to carefully negotiate the confusion of trackwork that snakes the train into platform 12.

Carefully and sedately, the train enters the platform and comes smoothly to a stand with the engine about 10ft in front of the hydraulic buffer stops.

The platform is already fairly crowded with hotel porters, tourist reps and a large group of railwaymen waiting to unload the luggage vans. As the train rumbles into the station, newspapers are furled, cigarettes snubbed out and little card placards with the names of intended clients held so that the train passengers can see them. Behind a pile of large packing cases, out of sight of the platform inspector, a card school is quickly brought to a conclusion and the railway employees move towards the point where the luggage vans will stop, ready to offload that cargo into a series of lettered metal cages, according to the name on the baggage labels.

Most arrivals at Waterloo are frenetic affairs with passengers disembarking on the run, well before the train has stopped. The first passengers are usually through the ticket barrier while the train is still moving. Boat train arrivals though are more leisurely occasions.

During the ensuing period where passengers are reunited with their own small mountains of luggage, hotel porters and travel agents establish contact with their allocated guests, friends and relatives greet each other, the engine is uncoupled from the train and both crew members quietly observe the activity from the relative calm of their engine.

Suddenly a group of tourists break away from the organised chaos on the platform and point their ever-ready batteries of cameras at the engine and crew. With words of wonderment and appreciation that such a comparatively small machine could have put up such a performance, they click happily away taking their pictures for the folks at home.

At the far end of the train, a tank engine enters the platform and couples on to the train. Although many of the travellers have departed to their respective destinations, there are still quite a few milling around as the tank engine is given the road to take the empty stock back to Clapham. The Nine Elms crew give two whistles, which is answered by a cockcrow from the train engine that proceeds to start pushing the train out of the platform. Only when the train has started moving will the tank engine, now at the head of the moving train, start to power the empty carriages back out of the terminus. Having given the London crew a goodly shove, the train engine eases back and allows the train to pull away before coming to a stop in front of the signal that has returned to the 'on' position at the end of the platform. After a few minutes' wait the signal is again pulled off to allow the engine to canter back to Nine Elms engine shed at Vauxhall.

Once in the sanctuary of the shed, the fire can be cleaned, smokebox and ashpan emptied, coal and water replenished and the engine turned ready for a run back to the south. When all is done the engine can be left, and the crew can visit the enginemen's mess room for a good wash, a can of tea and meal break in jovial company, followed by more rivalry in the company of their Cockney colleagues.

No. 34056 *Croydon* speeds down towards Winchfield and under the high bridge known as 'suicide bridge'.

6
Work and Play

Footplate work was hard, hot and very responsible, requiring at all times maximum effort and concentration. During the time spent actually working on the main line, there was little scope for frivolity. That is not to say that all was seriously dull. Away from the public face, particularly in the closed environment of an engine shed, the play and practical jokes could be equally as hard as the work itself.

As a rule, the pairing of regular drivers and firemen was an acceptable working environment and indeed sometimes a happy state. Whatever the differences in characters of the two men, they had to spend long and often hard hours working together as a team. There were, of course, exceptions to that rule. There were a very small number of the most senior drivers who had joined the railway in the golden Edwardian age of the London and South Western when the driver was a god-like figure, and their firemen perpetuated that status throughout their own driving career.

One of that small band of martinets was an Eastleigh man who spent his time being as awkward and hard on any fireman that came his way as he knew how to be. This driver, who we shall only call Reggie, was positively loathed by all who had the misfortune to be his fireman at any time. By the early 1960s, many of those who harboured grudges were themselves drivers and no longer walked in fear of being reported for any misdeed – actual or contrived.

It was also at this time that Reggie, being very near to retirement, opted to leave the main line and spend his last couple of years quietly as a shed engine driver. His work involved moving engines within the confines of the engine shed with the assistance of a very junior fireman. It actually meant he could spend his time doing exactly whatever he wanted and leave all the work to his young assistant, who would never dare make complaint.

A study of Eastleigh station area with the third rail laid beside the track ready for installation. The engine on the Bournemouth service is No. 75074. This is the same type of engine that collided with a tree on the Botley line in March 1967.

It happened one morning when Reggie was on early turn; the busiest period with all the night freight engines coming on shed, that one of his former firemen, now a driver, was also about the shed complex preparing engines for their day's work.

While so engaged, our nameless driver observed Reggie slope off with a newspaper under his arm towards a certain building and remained there for a good half hour, while his fireman and the other pair of men on early turn worked frantically to keep the coal stage clear of incoming engines. As he was on the same duty all week, our driver decided that it was retribution time for the hated Reggie.

During the course of his work, the assailant 'borrowed' some life-expired detonators from the bin at the back of the stores and carefully prised them open to reveal their gunpowder content, which was then emptied into a cardboard tube that once contained a well-known brand of sweets. The powder was carefully packed down with cotton wool waste and a short fuse inserted. With this king-sized banger concealed in his jacket pocket, the deliverer of retribution awaited his opportunity.

Between the very busy area of the coaling stage and the back of the equally busy engine shed, there was a long low building housing several cubicles. Placed in such a position, the building was subject to clouds of coal dust from the coaling stage on the one side and black sulphurous smoke from idle engines on the other. Although when newly built this lavatory block was a pristine red brick and tiled roof structure, the intervening decades had seen it heavily weathered and encrusted in soot and coal dust.

Almost on cue, Reggie was seen entering his favourite cubicle with the morning paper under his arm.

Being a building designed for a specific purpose, the lower edge of the cubicle doors were in fact some 6in above the concrete floor to allow a certain ventilation and, after allowing a period of time for Reggie to settle to his task, a clandestine figure sneaked up to the closed and bolted door, lit the fuse and rolled the ordnance with sufficient force to ensure it gained the back of the cubicle and was not, therefore, immediately obvious to the occupant. His task achieved, a hasty retreat to the cab of a nearby engine gave a good vantage point to observe the action.

It is difficult to state the exact sequence of events that followed, as the action was fast moving. The initial stage was something of a thunderclap, which caused the windows, doors and roof tiles to shake. This in turn release several decades of captured soot, coal dust and general grime from the fabric of the building. The net result was a dense cloud of black dust intermingled with smoke from the firework, a loud scream and several words of discouragement from within before the cubicle door burst open. Out of this conflagration, now billowing from the open cubicle, came the blackened figure of a man hopping with both feet; his trousers round his ankles and his now very grey-looking long johns being desperately pulled up to his waist, while remnants of the now abandoned morning paper fluttered in the wind.

The reaction by other shed staff to such a bang would only cause those in the immediate vicinity to look up and witness a very agitated Reggie trying to retrieve his trousers and what little that remained of his pride. In the heavy engineering environment of a locomotive running shed with constant maintenance and servicing, loud noises and bangs were commonplace, albeit the toilet block was a more unusual source.

Indeed, the whole railway was a fairly noisy affair. For audible warning in emergency situations, very loud and distinct signals were devised to be heard over and above any conflicting sound. So distinctive and rarely used were these signals that they immediately conveyed their message to all railway staff.

For a person at the line side to attract the attention of the driver of a fast-moving and noisy locomotive, three detonators were place on the rail at 10-yard intervals. The explosion of three detonators would trigger an immediate reaction by the enginemen to cause the emergency brakes to be applied and steam shut off. In the language of the rule book, which all enginemen knew and carried as part of their personal equipment, three detonations means 'obstruction: danger' and struck fear into the heart of even the hardest of enginemen.

For engine crews to draw the attention of others, a series of 'pop' whistles would be given. This was the sounding of the whistle in short, sharp, blasts continuously and meant the train/engine was running out of control.

At the side of Eastleigh shed, adjacent to the office block, was a single-storey, brick-built classroom. This was in fact the firing school. Within these walls, young men who had spent their time as cleaners were given theoretical instruction on the working and function of the steam locomotive. This schooling was designed to collate the knowledge already acquired in the course of their duty and was aided by many beautifully engineered, working models of the various parts and motion.

Apart from intricate knowledge of the locomotive and best practice in firing techniques, a major part of the curriculum was the rulebook. To pass as a fireman, the student had to demonstrate a sound knowledge of the many rules that governed the safe working of a locomotive and train, but to a young man wanting to understand the inner working of the steam locomotive, such rules and regulations were tiresome and boring.

On a warm summer afternoon, with the smells and sounds of the steam railway not 10 yards from their prison window, the reluctant students found it impossible to give the matter their fullest attention. A practical demonstration, out in the fresh air, was the answer. So it was that the instructor led his little band of students over to the south of the running shed and assembled them in a group to observe and take notes.

With his charges arranged to get the best view, the instructor proceeded to lay three detonators, at the prescribed intervals, outside the shed on No. 8 road; the through road. Taking the most available engine stabled at the back of the shed, he then proceeded to drive it up No. 8 towards the shed and the waiting detonators.

Bang – bang – bang was clearly heard by everyone, both indoors and out, for up to about a mile away. Onwards the engine went, towards the shed, where just before entering the instructor driving started to sound a long series of pop whistles until the engine was well inside the shed.

Having given his demonstration, the engine was quickly reversed and put back from whence it originated. The effect was somewhat dramatic.

Initially the three detonations brought fear and consternation into the hearts of every enginemen peacefully engaged upon his orderly duty about the shed. Oil feeders were dropped, fire irons abandoned, as footplatemen hurried toward the unseen calamity to render assistance. Such noble actions were immediately reversed with the even more spine-chilling sound of pop whistles, indicating that a train or engine was running out of control. Engine crews and shed staff all scattered to find whatever safe cover they could to await the inevitable crash, not knowing where, and into what, that crash might be.

By the time it was realised that there was to be no crash, a sizeable crowd of would-be assistance had gathered to witness, not carnage and devastation, but a medium-sized tank engine quietly making its way back along No. 8 road, while a band of trainee firemen sauntered back towards the classroom having had a very visual lesson in emergency procedures.

Questions were asked at the highest level; interviews between shed master and instructor ensured that henceforth and forever more, under no circumstances would a repeat performance be tolerated and if such a graphic lesson was needed in future it had better be conducted at the far end of the shed yard where the only creatures to be panicked were the rabbits in the fields, and then only after having given due notice.

The rule most commonly used rule during the daily routine of enginemen was rule 55; a lengthy instruction detailing the action to be taken when a train is brought to an unscheduled stop. The scenarios covered were anything from a full-scale emergency to the routine stop at an adverse signal. Modern colour light signalling has rendered much of that rule obsolete, but in the era of semaphore signals it was very much a part of life. Broadly speaking, when a train stops at a signal, it is the responsibility of the engine crew to remind the signalman of its presence. Very often a signal is equipped with a device called track circuit, which gives a visual display by means of a red light on the signal box track diagram. Many signals closer to a box were also equipped with a telephone link to the box, so you could report your position by phone. However, there were also still a number of signals that were neither track circuited or wired for telephone links and these tended to be more outlying signals where it was not anticipated that a train would need to be stopped under normal circumstances.

In these situations one of the engine crew was required to walk to the signal box and physically report the presence of the train, sign the train register to state the reason for their visit and stay in the box as a reminder to the signalman, until he could give clearance for their train to proceed. Just occasionally circumstances dictated that such a stop would be made and although it is not written into the rule book, experience has shown that part of the criteria for stopping a train in such a situation would appear to be that it must be a dark night with the most inclement of weather.

One of those occasions that remains within my memory happened on a bitterly cold January day in the very early hours. This particular week's roster was to work the early freight from Eastleigh to Bournemouth. Not an especially arduous task and on this particular day all was well and we were given a clear run all the way through Southampton, out to Brockenhurst; up the hill to Sway and New Milton and on to Hinton Admiral, where things suddenly changed. Christchurch's distant signal was on. This meant we had to prepare to stop at that station's outer home signal, which in the event, is exactly what we had to do. The outer home signal is the first stop signal protecting a station and at Christchurch it is probably about three-quarters of a mile from the signal box, perched up on an exposed embankment neither track circuited nor connected by phone link. Ordinarily, unless Christchurch had a problem, there was no reason to be stopped there.

The enclosed cab of the Standard locomotive we were on was warm with the fire doors thrown wide open against the unexpected cessation of the demand for steam. Outside it was bitterly cold and I was well aware that, unless we were given the road in the next few minutes, someone was in for a long and cold walk.

It made no difference how much I glowered at the thing, or how hard I cursed it under my breath, the signal remained resolutely at danger, almost mocking the fact that it was the cause of our displeasure. As expected, my driver, who was by now standing in front of the fire warming his backside, suggested that it was time that I went for a little walk to wake the Bobby up, but to be sure I put on my reefer coat as it might be a little nippy for a long night walk.

Suitably attired against the weather, I descended the cab steps, only to have one foot slide away from me as I touched the ground. The other foot then also lost its grip on the footstep and scraped the skin on my shin as I alighted in an undignified manner, clutching the spare engine headlamp.

Engine headlamps are lit by paraffin and are designed to give a light that can be seen, but not to see by. Its purpose was to advertise my position and in no way illuminated my way as I stumbled along the cess beside the track. It would have been easier to walk along the track, but as the driver would not be able to see me directly in front of him and as the breeze was against me, I should not be able to hear the engine approach if the signal had cleared and my driver decided to pull gently ahead to pick me up. Therefore, for safety reasons, I was obliged to stumble along a frozen track high up on an embankment overlooking the water meadows beside the River Avon. A charming scene in the summer as we glided by en route to Bournemouth but definitely not charming on a dark winter's morning when you are walking. The further I progressed, the colder I became and I was convinced that my driver was entirely correct in the assumption that the only possible reason for our enforced stop was that the signalman had fallen asleep and was just waiting for me to politely knock on his door and request that he attend his duty!

I had plenty of time to consider my actions on arriving at the box and how someone was going to get a rude awakening. His train register book would certainly be annotated with the exact reason why I was on a mile hike and already frozen to the marrow.

Onwards I tramped, over the river bridge, under the road bridge and there, high up in a position that allowed a wakeful signalman to see the railway over the top of the road bridge, was my goal.

From ground level there seemed at least a hundred steps up to the box, each one frozen and covered in a white coating of frost. Still clutching my lamp in one hand and the stair rail in the other, I stumbled, slipped and slid up each one of those steps, whose whole purpose seemed to be to cause me as much difficulty as possible. My humour decreased accordingly.

The outer porch door was, as is usual practice, unlocked and allowed me to get out of the cold and put my lamp down. The inner door to the signal box itself would be locked against intruders at night and I steeled myself to virtually bang the door off its hinges to let the incumbent know that he had a very irate fireman at his door. Just before assaulting this barrier between my quarry and me, I noticed through the window that the signalman was not in fact asleep but sitting quietly in his armchair with a cup of tea at his side, reading.

Sensing my presence, he immediately jumped up and hurried to unlock his door. 'Come in, come in buddy, get yourself warmed up. God, it's not the sort of night you want to be out walking the line. I'm so sorry, but it can't be helped – cup of tea? Totally bemused and horribly deflated, I could only reply with, 'Yes please, what's the problem then?'

He then related the situation that the track maintenance people had occupation of the up loop through Pokesdown station, which is the next station on the line. They were replacing rails and using a crane. Because the crane needed to swing out, it would foul the down through road and they had therefore requested a temporary halt to the traffic on the down line, just long enough to complete their manoeuvres. That was an hour since and they still hadn't given a clearance. There was another freight train stood at Pokesdown's inner home signal. A parcels train had stood in Christchurch station until the signalman had verbally instructed the driver to proceed with extreme caution towards Pokesdown outer home and we were stood out in the wilds at Christchurch's outer home; as regulations demanded. Now that the platform was clear and the preceding train protected by Christchurch's advanced starter signal, the signalman, in my presence, pulled the signal that we were held at, so that the train could draw forward to the next signal, which was the inner home.

All this took a while, by which time I had warmed up in front of his cosy fire, had the benefit of hot tea inside and the world didn't seem quite so bad.

Having remarked that I was now a slightly lighter shade of blue, my hard-pressed colleague suggested that he bring my train in, if I would kindly meet it on the platform and ask the driver to pull up to the platform starter signal and wait for Pokesdown to get themselves sorted out, making certain that I relayed the signalman's apologies to my driver. He also reminded me to sign the train register before I left.

All of those things I had intended to say; the remarks that were to be entered in the register, was now all a thing of the past. I simply entered which train we were; where we were held; the time I arrived at the signal box and in the 'Reason for visit' column, just noted the usual 'Rule 55', before signing the entry. With a cheery, 'Take care and mind those steps mate,' from the signalman, I buttoned up my overcoat, retrieved my lamp and left him to his solitude.

If anything, the steps down were worse than going up and I found myself stumbling and sliding back down to Mother Earth, clinging on to the handrail just to stay in a more or less upright position!

Having regained the warmth of my engine and enlightened my driver of the circumstances, it occurred to us that this situation posed a bit of a problem. Although we were booked a twenty-minute break before working our train home, that depended upon arriving in time to take that break. The alternative was to eat your meal while working your train. As we were being held up for an unspecified time, we decided that Christchurch station on a cold winter morning was not an ideal place to have our meal, but better than missing it altogether. As there was a signal box with a kettle full of hot water to make a can of tea just down the other end of the platform, it was in fact a very good place to have breakfast. However, as I had just suffered a long cold walk, I persuaded my driver that he might like to stretch has legs and be the tea maker. As he left the engine with tea can in hand, muffled against the cold, I completely forgot to mention the perilous state of the long climb up the steps to the signal box ...

The period 1965–66 was one of intense engineering activity throughout the whole line from Woking to Bournemouth. This was due to the installation of the infrastructure in connection with the extension of the third rail electrification. This major work involved the addition of the third rail for current collection as well as buildings for electrical generation and switching. Allied with this modernisation, a programme of colour light signalling installation between Southampton and Woking saw the commissioning of new power signal boxes at Eastleigh, Basingstoke and Surbiton.

Initially much of the work was done at the weekends when traffic was lighter. Wherever possible, trains would be diverted from their normal route to allow the engineers to have complete occupation of a section of line. Many Bournemouth trains were diverted at Woking to take either the Portsmouth line to Cosham and back to their correct route via Fareham and Netley to Southampton, or take the Alton road to get to Winchester via Medstead. This last route was a secondary, single-line track with severe gradients calling for skilful and hard work by the enginemen when they had thirteen well-filled coaches behind them.

Southampton, too, could be bypassed but only by the by the long-winded task of attaching an engine to the rear of the train at Eastleigh, then pulling it essentially backwards to Romsey. Now it would go forwards again to reach Southampton via Nursling and Redbridge. At Southampton the train engine detached and the ensemble proceeded to Bournemouth (now in the reverse direction of travel) by the engine attached at

Eastleigh. Naturally this caused heavy delays to the timetable and was only initiated during a couple of weekends when single-line working was not an alternative for whatever reason.

Between Southampton through to Bournemouth there was, at that time, no alternative route. Previously a secondary route from Brockenhurst via Ringwood and Wimborne to Bournemouth existed; in fact it was the original main line. However, this was closed and the track lifted in 1964. Those of us who served the railway at that time found it incomprehensible how the planners of these matters could lift the track in the knowledge that a major feat of engineering was about to commence that would seriously affect the main arterial route to Bournemouth with its heavy concentration of holiday traffic.

Thus most of the electrification work between Southampton and Bournemouth was done during the weekday with single-line working and on Sundays when full closure of sections of line resulted in a shuttle bus service to bypass the affected section. None of this, of course, went down very well with the travelling public, or the railway operating staff who had to try and extract some professionalism out of a very unsatisfactory situation.

Single-line working involves a different set of working practices for both engine crew and signalmen. The single-line Block Token system, whether permanent or temporary, allows the electrical locking of signals and pointwork by the mechanical operation of turning and withdrawing a key from a machine. Once the route has been set for a particular train and the key withdrawn, it cannot be unlocked or the route altered until the key is inserted into a corresponding machine at the far end. There were, of course, emergency override procedures, but essentially the key was given to the driver of the train for which the route was set and all the time that key was in his procession the driver knew he alone would be travelling on that section. It was not possible for the signalman at either end to inadvertently send another train into the same section.

The key, or tablet was a robust brass item that, when withdrawn from its machine, was placed in a stout leather pouch with a large, leather-bound steel hoop attached. The method of transferring the token from signalman to engine crew, without the train stopping, was for the signalman to hold the tablet lightly by the pouch and hold it up at arm's length so that the hoop was extended towards the engine. The fireman would lean out of the cab with his left arm extended and aim to put that arm through the hoop. When the hoop had made contact over the fireman's arm, it would be snatched out of the grasp of the signalman. Once back inside the cab, the fireman would read aloud the section description on the tablet so that both driver and fireman were satisfied they had the right one and then hang the hoop over a convenient control valve in full view of the driver. Under no circumstances would a train proceed without the correct tablet in the safe custody of the engine crew.

For setting the tablet down at the far end, it was either handed back to the signalman in a similar manner, or as was usual with a permanent single-line arrangement, a set-down post with a padded arm parallel with the track and about 4ft distant allowed the fireman to lean out and thread the hoop over the arm. If he failed, he would throw the tablet onto the ground near the post.

Complications arose when travelling from one single-line section directly on to another, for at that point the old tablet had to be set down at the same time as the new tablet was collected.

On the engine, the fireman needed to hold out his left arm to receive the new tablet, while at the same time holding the discarded one in his right hand with the hoop extended down towards the waiting signalman. Thus, with both hands in use, the fireman had nothing to maintain a balance on the engine and had to rely on the waist-high cab door, if fitted, or the safety chain across the cab in a more elderly engine. A good sense of balance was required with the engine lurching over the inevitable pointwork. The maximum regulation speed for these operations was 15mph, but a good fireman would pick up or set down at well over twice that speed, although a full changeover would necessitate a reduction to around 20mph.

While the enginemen were well versed in such operation due to their experience on single-track lines, signalmen used to operating in main-line boxes were not and it is true to say that some men used this disparity of experience for a source of their amusement.

The section between Beaulieu Road station and Brockenhurst is a fairly long one cutting right across the New Forest and this was worked as a single-line section for three or four weeks continuously as the third rail was laid.

Beaulieu Road station is situated some miles from any reasonable habitation, in the heart of the forest, and saw few trains stopping and probably even fewer passengers outside the holiday season. The signal box there served the function of an intermediate box on the otherwise exceedingly long section from Lyndhurst Road to Brockenhurst. Thus the normal work of the signalmen was merely to pass trains from one section to the next, with about five signals to operate in each direction.

Suddenly though they were now faced with the prospect of having a single-line machine installed that electrically locked their signals controlling the singled line, and the previously unknown experience of handing the token in its pouch and hoop to non-stop trains.

As the signal box was not designed for such operations, the signalmen needed to stand on the station platform at a distance of about 4ft from the edge while the incoming train coasted towards them. To anyone not used to standing close to a moving, hot, hissing, steam engine, this itself can be an unnerving experience, even at slow speed.

It happened that during this period my driver and I were working a Bournemouth semi-fast on one of the weeks when the former down-line only was being used as the running line. This meant that in the down, or Bournemouth, direction there was no necessity to change lines and progress was only impeded by the necessity to reduce speed to pick up the tablet.

On the first day we coasted, fairly sedately, through the platform to espy the signalman standing as he was instructed to do, with the tablet held above his head, but obviously with some trepidation.

He was clearly very nervous about the whole thing and instead of the hoop being held steady for me to aim at, it appeared to be wandering from side to side. The closer we approached it became apparent that the tablet holder was in fact standing with his eyes shut and face contorted as though something dreadful was expected. Having managed to grab the hoop in my arm on one of its closer wanderings, I was almost pulled off the footplate as the deliverer, instead of lightly holding the pouch, was grasping it for dear life!

Naturally the driver had noticed these things as well as I and with that mischievous grin, culminating at the end in the ever-present cigarette, he observed, 'The bobby's a bit nervy ain't he mate.' It was clear that he intended to have a bit of sport the rest of the week!

On the following days we approached at increasingly greater speeds and each time it was touch and go whether the proffered hoop would be close enough for me to catch. I found myself leaning further and further out of the cab to reach as the signalman edged further away. On the penultimate day, just to add to the effect, the driver touched the whistle as we entered the platform, causing the hapless fellow awaiting our arrival to take another half step away from the platform edge. Rather than catch the hoop in my arm, I just about managed to grab it as it hit the palm of my hand. On the final day of our turn, we approached the platform at a good 35mph with the whistle blowing. Halfway along the platform, the engine was opened up vigorously so that by the time we reached the pick-up point we were travelling at a reasonable speed with the engine working hard. That was just too much for our nervous friend.

He took several steps away and let the hoop drop down. By this time I was hanging out the side of the cab as far as I could with my right hand holding onto the cab handrail, trying to gauge where the hoop would be at the point of contact. Feeling the hoop strike the tops of my fingers, I immediately curled them round to gain some purchase. Unfortunately, the other end was still being held fairly tightly, with the result that it was wrenched out of his grasp and mine at the same time.

The thing span up into the air and landed halfway down the roof of the first carriage, before lazily sliding to the ground beyond the platform end.

We had no choice but to make an emergency brake application as the instrument of our safety was now lying back along the track. Although I dismounted the engine before the train had come to a stand, there was still a good quarter of a mile run back to retrieve the tablet, which was achieved in full view of the signalman waving his fist, doubting our parentage and threatening to report us to the highest authority!

One of the more pleasurable tasks in the Bournemouth Link involved working the 13.30 from Eastleigh to our usual destination. A very civilised signing on time of 12.00 was allowed to prepare the engine before collecting the train from the carriage sidings and working non-stop to Southampton, where we were to connect with a Portsmouth–Salisbury train, thence all stations to Bournemouth. As the engine diagram called for this engine to work to London and back after we had taken her to Bournemouth, the turn was always booked for a big engine, usually of the West Country class. For the five-coach 13.30, this was excessive overpower and my preferred method of working was to build up a good heavy fire between Eastleigh and Southampton, thus allowing me a relatively easy trip to the seaside. Apart from the occasional round of fuel to keep the fire in shape, my main occupation was to ensure the boiler levels were kept sufficient to counteract the explosive departures and emergency-type stops that were the characteristics of my driver's preferred method of driving – on any train.

Meanwhile, at Eastleigh the buffet on the down-side station platform had, at that time, some rather tasty steak and kidney pies. It became customary on this particular turn for either my driver or I to go to the buffet after we had reversed the carriages into what was then platform three and buy two of these pies. Buying them cold was also the cheapest option and as we would not have time to eat them until we arrived at our destination, we would wedge the two pies on top of the boiler backhead by the steam fountain. There they would stay safely, getting nicely heated through until we had a break, whereupon they would be piping hot, giving a wonderful aroma and tasting delicious.

On this particular day my driver elected to go to the buffet and on his return announced that they only had one pie left. As he was unaware of my other tastes, he didn't actually buy anything for me. By this time it was far too late for me to go wandering off, we had a train to work. However, having only one pie presented a problem as it could not be wedged so securely in the normal spot where two fitted snuggly. Eventually though he satisfied himself that all would be well and I enquired whether he was intending to share his meal. His response was to the effect that half a pie would satisfy neither of us and it was far better that he had the whole pie because then there would only be one unsatisfied person rather than two – logic I suppose that is hard to fault.

As usual, we left Eastleigh as though the devil himself was chasing us and my attention was quickly diverted to the sole occupation of building up my fire. Round after round went into the firebox; around the back corners, down the sides and under the door. As we rattled through St Deny's I had nearly done, just a few final shovel loads down the middle.

With the loaded shovel in full flight, I was vaguely aware of something come down past my head, strike the shovel and disappear into the fire along with the coal. It was obviously not very heavy as it in no way deflected my aim, neither did I feel any additional load on the shovel.

Slightly perplexed, I checked the boiler controls and all was well; clearly nothing had fallen off. A check inside the fire revealed no metallic foreign bodies, just burning fuel, so clearly I was imagining things.

As we screeched round Northam curve, past the dingy house backs, through the tunnel and into platform four at Southampton Central, my attention was with my work rather than what may or may not have gone into the fire.

With a few minutes to wait for the diesel unit to come in behind us and stop at platform three, my driver, who was quite short, left his seat and stretched up to check his pie, while I busied myself sweeping up the coal dust from the cab floor.

Presently there was a thump on my shoulder and, somewhat startled, I looked round to see my driver with both hands on his hips. 'All right, very funny. Jokes over, what have you done with it?' he demanded.

'I don't know, what are you going on about.'

'You know what I'm talking about, my bloody pie, where is it?'

'Where you put it, I expect,' and as I said those words I suddenly realised the source of my mystery object. Immediately the funny side of the whole occurrence struck me and I quite involuntarily dissolved into a state of hysterical laughter. So much so that I was obliged to perch myself on my seat and lean against the cab side for fear of falling down.

Through bouts of uncontrollable giggles, I managed to relate the happenings just after St Deny's, but without much conviction. My mate could not accept the coincidence that his pie should fall down at the exact time to allow it to converge with the flight of my shovel and conveniently pop into the firebox along with the coal. In his view it was far more likely, given our earlier conversation, that I had either deliberately took down his pie and disposed of it, or at the very most, it fell of its own accord on to the cab floor, where I had seized the opportunity to throw it in the box in a fit of petty jealousy.

As I was still very much amused by the episode and, in his view, not taking this as seriously as the occasion demanded, I did nothing to convince him otherwise. Taking umbrage, he sulked over to his side of the engine, enquiring to himself what he had ever done to deserve such a mate as me. I do believe, to his dying day, he firmly believed that I deliberately deprived him of that pie. Very little conversation took place between us on the run to Bournemouth, but the engine was being driven even harder than usual.

Having pushed the carriage set back into the siding at Bournemouth, it would have been the time to partake of our hot pies. Instead, I had a driver who was clearly ignoring me in deep offence.

'Now look Dave,' I said. 'I didn't put your pie in the box on purpose. If it will make up for it I'll buy you a pint when we get back.'

'So you think a lousy pint will make up for the loss of my pie do you?' he enquired. 'What about the loss of the pleasure in eating it, you've got to take that into consideration'

He stood up and thrust his hands in his serge jacket pockets 'Try me on two pints,' he said.

'OK two then, if it'll make you happy.'

'Yes,' he said, 'I think that might do the trick. In fact, I feel a bit better already.'

His countenance erupted into his broad grin, almost from ear to ear, and always with a Player's Weight cigarette parked at one end, 'Oh, and yer not such a bad mate – for a young un,' was probably the best compliment he ever paid any of his firemen.

Many years after I had promotion above the Bournemouth Link, indeed long after the demise of steam operation, my old driver and his secondman were booked to work a petrol train from Fawley to Birmingham, being relieved by Western men at Banbury.

The usual practice then was also to have a meal break in the old driver's cabin of the long-defunct engine shed. By this time the intense and at times bitter rivalry between Southern and Western men had subsided. They were all now part of the newly modernised and uniform railway. They all wore the same 'diesel green' uniform and there were no longer any distinguishing badges. It was then quite normal practice for 'foreigners' to mix and take their break together.

Having entered the driver's cabin in the small hours, they were the only crew present and proceeded to make their tea and open their lunch boxes. Almost as though he had been invited, a mouse appeared from somewhere and, as bold as you like, sat beside the table waiting to be given some titbits.

Now, driver Batten may have been a terror to fire to, but he always insisted on a tidy footplate to work on and equally took exception to eating his meal in squalid surroundings. To have his meal break interrupted in such a way was proof that this was a vermin-infested hole and not fit for decent enginemen, even if it was all right for Western men. With flailing legs and arms accompanied by liberal amounts of shooing and chasing, the mouse was persuaded to disappear. The following morning both men again made their way to the cabin for their break.

With tea made and lunch boxes at the ready they were again joined by the same mouse, or at least its double, who hopped up on to the bench beside the dining driver. He did not stay very long and amid much cursing and swearing, beat a hasty retreat closely followed by anything and everything that was throwable.

The third night driver Batten was ready. Before entering the cabin, he armed himself with a stout piece of timber, which he placed by his side as he took his seat. Sure enough, as soon as tea was ready, out popped their uninvited guest.

Cunningly, a piece of sandwich was thrown onto the floor where the mouse sat. As soon as his back was turned to devour this morsel, the makeshift cudgel was brought to bear and with a mighty 'thwack' it killed the little fellow stone dead.

Before leaving to resume their duty, the body was placed on a newspaper and left with a note stating how disgusting the cabin was and that now a start had been made to clear the vermin, perhaps Banbury men might like to continue with the work to make their cabin habitable.

The following day, expecting to find some note of thanks, they again proceeded to the cabin. The note that was left made no mention of thanks. It stated that the mouse had been buried with due honour in a plot at the rear of the old engine shed and that if the Eastleigh men who perpetuated the crime were still around when the next set of Banbury men were in the vicinity, then they would also be whacked over the head and buried in a similar spot! Judging by the liberal use of profanities within the text, it seemed clear that this was no idle threat.

Fearing the resurgence of old rivalries for some unexplained reason, the Eastleigh men considered it prudent to take their can of tea and retreat to the station, where they had their meal on a platform seat – as indeed they did the last night of their turn on that duty.

The next week a different pair of men worked the petrol train. They were, naturally, unaware of the events of the previous week, or of how the tale of the dastardly deed had by they reached the ears of every Banbury engineman and men from adjacent sheds. Their own quiet meal break was rudely interrupted by a delegation of men from Banbury shed and they found themselves the subject of an inquisition demanding to know the identity of the detestable 'Southern' men who had murdered their pet mouse!

On Thursday, 9 March 1967 we worked a freight from Eastleigh to Northam, then a further freight from there to Portsmouth on what was a very wet and stormy night. Being in the Holiday Relief Link, I was paired with driver Cook for the week.

This particular gentleman originally came from and still lived at Salisbury, each day making the journey to and from Eastleigh in his well-groomed, highly polished Morris Minor car. The conclusion of this duty turn was to travel light engine from Portsmouth back to Eastleigh shed. Light engines, that is just the engine and no train, are in theory limited to travelling at 60mph due to a lack of brake power.

For Eastleigh men going home, particularly one that lives in Salisbury, such speed restrictions are open to interpretation. Our engine for the day was Standard class 4 tender engine number 75074.

Having cleared the curve into Fareham station, we accelerated away, over the deviation to Knowle and on towards Botley. Outside it was pitch dark and wild. Under clear signals, we raced through Botley, giving the customary pip on the whistle to the signalman and out into the dark night again.

I had the fire doors about half open, which meant the cab was illuminated with the light of the fire. Because we both had our cab windows drawn shut against the weather, we were looking ahead through the front spectacle glass.

Approximately midway between Botley and Eastleigh South there was a sudden loud crash and bang from the front of the engine, which lurched violently. This was followed by a screeching, tearing, scraping noise.

Our actions were swift and automatic. My driver shut off steam and wrenched back the brake handle for a full emergency application, while I kicked the fire door shut and in one continuous action put on the injector. If anything untoward is about to happen, the last thing you want is the fire door open. These actions took but a few brief seconds, by which time the noise had stopped.

I vividly recall holding on to the injector steam valve and bracing myself against the cab side, waiting to see which way she would roll over. I am not entirely sure what I thought I was going to do when I did find out, but for what seemed like an eternity all noise had stopped. There was complete silence, no engine noises, nothing. Somewhere over the other side of the cab, my driver was also stood at his controls, waiting. There was no heroic action beloved of movie-makers. No recital of prayers, no quick verse of 'Abide with me'. We just waited to see what would happen.

Suddenly, after a couple of seconds that seemed like days, I became aware that the normal noise of a fast-moving engine had returned. The sound of wheels on rails, of cast iron brake blocks pressed hard against steel tyres; and we were rapidly slowing down. The engine came to a stand – still upright – and we both stood on our respective sides, motionless for several seconds, taking in what had just occurred.

Pushing the fire doors open again revealed a cab full of debris. Branches, twigs and leaves filled much of the cab and I was conscious of a cloud of dust that was drawn into the firebox through the open doors. Although the cab windows were shut, all this debris had come in through the gap above the cab doors between engine and tender.

I think it was my driver who broke the stunned silence when he let out a long low whistle and very slowly and deliberately exclaimed, 'Bloody hell mate.' Our first priority was get down and check the engine, which entailed clearing away a pile of branches from the cab door before the driver turned to climb down backwards off the engine. He stopped with one foot dangling over the cab side trying to find a foothold, which was no longer there. We instead alighted from my side.

In the dark we could make out that my side of the engine was festooned in bits of tree and foliage. Walking around the front revealed that we were no longer carrying any lights. Two of the three lamp irons over the front buffer beam were almost flattened against the framing and the lamp that had been placed on one of the irons was completely missing, while the other lamp up on the smokebox was a flattened mass. The nearside buffer was pointing skyward at an angle of several degrees, while the casing on the nearside cylinder had been crumpled and torn away like a piece of paper to reveal the cylinder casting and insulation underneath. The missing footstep under the nearside cab was in fact bent right up under the tender frame.

As far as could be seen in a black night with a gale lashing our faces, everything else *seemed* to be in order and at least all wheels were still on the track. Carefully steam was applied and the engine moved slowly forward. All seemed well and we decided out best and quickest option was to proceed with caution to Eastleigh South signal box. It has to be said that the engine's exhaust was making a very peculiar noise, but I assumed this was due to it being disguised as a tree.

Stopping outside the signal box, we summoned the signalman's attention and reported that we had hit something on the line, requesting we be let into the engine shed as fast as possible. From the light of his box he could see better than us the state of our engine and without further ado he waved us on and could be seen urgently sending the six beats on the bell 'Obstruction Danger' message on back to Botley.

We limped on up to Eastleigh station and in the few minutes it took to get there the telephone lines had been busy. Hardly had we come to a stand in the platform when the points snapped over and Eastleigh West Box gave clearance back into the shed.

It was a fair delegation that came to meet us. The duty foreman together with several of our footplate colleagues came to see that we were not injured, which neither of us were. Under the floodlights of the disposal road we could, for the first time, see the extent of damage.

Superb view of a clean No. 34034 *Honiton* before she degraded into the filthy and unkempt state just before the end of steam. She is on Eastleigh shed being prepared for working an inter-regional service.

Apart from the wreckage we had noted in the dark, we could now see that the reason for the buffers' upward position was due to the fact that the entire buffer beam of 1in-thick steel plate was itself bent upwards. The smokebox door was stoved in. Down the side of the boiler cladding was a deep gouge and one of the boiler bands that held the casing in place had been sprung so that it waved around like an aerial through the remains of the tree festooning the boiler. Something had shattered the driver's side spectacle glass, although the crazed glass remained in situ. Another deep gouge was present on the driver's side cab, right through the engine number as though something had tried to cross it out, and there were three further gouges along the side of the tender. The footstep that my driver had tried to use was in fact bent right back under the tender framing. Having ascertained that we were uninjured, another pair of men took our engine and threw the fire out. She wasn't going anywhere for a while.

Our task, or at least my driver's task, was to write a report on the accident, which I had to sign. I noted that in his report the driver only showed our speed to be at about 60mph when we struck what we now knew to be a tree of the horse chestnut variety.

It was not until I arrived home, well after the time I should have, that I had chance to sit down and reflect. Then the full shock of the night's events struck home. In that dark night it was impossible to see what we had hit or indeed just how many inches we were away from death. Had the tree fallen with more of its bough across the line, we would not have stood a chance of survival. Judging by the damage sustained by the engine, it was a close-run thing.

Another incident that threatened my career, although of a different nature, occurred while working an up London semi-fast. We travelled as passengers from Eastleigh to Southampton Central to relieve Weymouth men. Our train was to call at Winchester and Basingstoke, then run fast to the capital.

As was normal practice, we firemen made the verbal handover as we took water, while the drivers did the same as they changed places on the footplate. My Weymouth colleague informed me that the tender was full of coal dust and rubbish, but the engine seemed to burn it and steam well provided you didn't try to put too much on at any time.

From my position atop the tender, holding in the water pipe, I did not need to be told the state of affairs in the tender; I could see for myself. There was hardly a lump of coal in sight; it was all small coal and dust! The exception was a massive boulder, about half the size of a man, perched at the back of the coal space. As the engine was apparently steaming well and I did not relish having to spend valuable time breaking up this lump or the possibility that it might jam in the shovel plate, I decided that it could stay where it was. At least I should have some lumpy coal for when the fire had to be cleaned at the London end, where I would also have time to break it into useable pieces.

The trip itself was not out of the ordinary and the engine steamed well enough provided the fire was fed continuously in small doses not allowing the dust to clog. What I had failed to take into account was the fact that while using the coal from underneath, the boulder perched across the back was becoming progressively unstable. With a modern, enclosed cab, the coal in the tender is not readily in the view of the fireman.

All seemed well until the usual high-speed dash through Woking. The notoriously rough junction with the Portsmouth Direct line always makes the engine kick, followed by the tender lurching in the opposite direction. As usual, Woking platform was crowded with passengers awaiting their train to London. Just as we entered the platform there was an explosion on the ground behind me and I immediately looked back to see what it was. All I could see was coal shrapnel spraying up the platform. My boulder of coal had been dislodged, fallen over the side of the tender and at a speed well in excess of 80mph had hit the platform ramp. A thousand pieces of coal were thus projected upwards as shrapnel off the ramp and into the waiting passengers.

Travelling fast through the station, I did not have time to witness the aftermath of this accident, but visualised dozens of people being mown down critically injured if not killed. The driver, being on the opposite side of the engine, was totally unaware of anything untoward; however, as he has the ultimate responsibility for anything that happens to his engine, I was obliged to inform him. His exact reaction is not printable, but he rightly observed that there was nothing we could do about it, but that we should be ready to expect a reception when we arrived at Waterloo.

For the remainder of the journey I had visions of a fleet of ambulances ferrying the injured away from Woking station and it was with considerable trepidation that I looked for a delegation waiting at the end of the platform as we ran into the terminus.

I fully expected to see uniformed police waiting to cart me off on manslaughter charges. Instead there was but one figure, a short, stocky person in a cloth cap; the unmistakable appearance of the locomotive foreman attached to Waterloo station.

As soon we stopped he informed me that we had been reported for a serious incident at Woking concerning spillage of coal from the engine. He made no mention of dead or injured people and I certainly was not going to enquire too deeply. As his post was an extension of Nine Elms jurisdiction, he had no authority over Eastleigh men, but demanding to know my name, he informed me that I was to be interviewed by my own shed master at Eastleigh before I would be permitted to sign on the following day. As the instruction was only for me, my driver cheered up considerably.

The return journey, sometime later, involved stopping at Woking on the down through line, so that the platform was again on my side.

Alighting from the engine, I went to survey the scene, imagining chalk outlines where the dead had fallen – as in all the best detective films – together with blood from the injured, or at least signs of cleaning up. Instead there was actually nothing to see, other than a few pieces of coal that had escaped the porter's broom. I convinced myself that in the two or three hours since the incident it was natural that the hysteria had died down and the station staff would have cleaned away all traces so that other travellers would not be distressed. It did little to allay the feelings of guilt or my foreboding as to my fate – that had just been put off briefly until the following day.

Eventually the long day ended and I sought refuge in my home. Sleep did not come to me that night, just the horrible vision of dead and injured people pointing their fingers at me. How could I have been so stupid not to ensure my tender was trimmed and safe?

The eternal wait eventually passed and like a condemned man I put on clean overalls, buffed up my boots, put on a tie and made, what I thought, could well be my last journey to Eastleigh Motive Power Depot.

Arriving there an hour before I was due to sign on duty and keeping clear of the driver's lobby so that I would not see any of my erstwhile colleagues, I quietly stole into the entrance passage leading up to the first-floor offices and the inner sanctum of the shed master.

At the foot of those dark, wood-panelled Edwardian stairs, I stopped and composed myself for a few moments. Then, taking a deep breath, I ascended to my fate.

The first floor office was the workplace of various clerks, diagrammers and secretaries, all of whom were busy about their allotted tasks and probably took no notice of the interloper from downstairs, but to me it seemed that every one of them was looking at me.

Across the office was the door to the inner sanctum of the shed master, also known as the Guv'nor, the Old Man and various other unofficial titles. Guarding the door was a desk at which his secretary sat. None could pass without her intervention.

She was a woman, a bit older than myself and very attractive with her long dark hair; the not unwilling recipient of many admiring glances and quite a few wolf whistles from burly enginemen as she made her daily journey to and from the office.

This day her attraction eluded me as I approached the desk, hat in hand in suppliant mode.

'Hello, have you got lost?' was her greeting as I came near. I assured her that I was not lost and had to report to the Old Man before signing on. 'Oh dear, that sounds a bit serious,' she said, 'What name?'

'Fireman K. Dawe,' I replied as quietly as I could in case anyone else in the office overheard.

'Ah, that's you is it. Yes, we've had a bit of a report about you. It's here in my pile somewhere.'

And with that she made great play in looking through a pile of paperwork before extracting a printed sheet with a heading from the Chief Motive Power Inspector, Waterloo. 'Here we are, I think he's already read this once,' she exclaimed and read the report herself.

'Oh my word, I think we're in trouble, wouldn't like to be in your shoes,' she continued in an airy manner. 'Just wait here, I'll see if he can see you now,' and taking the report, she left her desk, knocked on The Door and entered without waiting for a response, then closed the door behind her.

For several minutes I waited in that office like an errant schoolboy waiting to see the headmaster. Presently she reappeared and clearly enjoying her moment of superiority, announced, 'He's busy and he'll see you when he's finished.' Then waggling a finger in the direction of two chairs placed outside the door, she instructed, 'Just wait – over there.'

I seemed to have sat there a long time, when from somewhere, a buzzer sounded and the red light beside the doorway changed to green. 'Yes, that's you,' came the voice from the desk.

I stood, approached The Door, and knocked. There was no answer, only the voice behind me commanding, 'I wouldn't keep him waiting, if I were you. Just go in.'

This was it then! With a pretty secretary watching your every move, it's time to be a man. I entered as boldly as I could into a typical Edwardian railway office with a huge, imposing, mahogany desk, overlooking the panorama of Eastleigh Locomotive Depot.

The Guv'nor was sat at his desk reading that report. Respectfully I stood before the desk and announced, 'Fireman Dawe, sir. I believe you wish to see me.' He didn't look up from his report, he just said, 'I know who you are,' in a low threatening voice that did not disguise his displeasure at our meeting.

Having finished reading the report, he placed it on the desk in front of him, pushed back into his chair and looking up at the ceiling moaned, 'Dear God. OK, let's have your version. What have you got to say for yourself Fireman..

I wasn't actually expecting to be asked that question, I thought it obvious what had happened. However, I explained about the rubbish coal in the tender and the one large lump at the back and how it dislodged due to the rough riding junction just before Woking platform. How I had seen the lump explode on the ground, but that we were travelling too fast to witness the aftermath.

He looked at me intently, then leaned forward, clasping his hands together on the desk in front of him. Then very quietly, slowly and deliberately he commenced his condemnation.

'Do you know just how much steam coal costs us nowadays, Fireman? Well I'll tell you – a bloody arm and a leg, that's how much, Everyone is intent on pricing steam out of existence in favour of modern forms of traction and we have to do the best we can with what we can get.'

His voice rising steadily, he continued, 'I've got staff out there in the office on the phone all day trying to secure reasonable supplies. OK some of it is rubbish sent down from closed depots up north; you've got to take the rough with the smooth.'

With his voice rising higher still, he reached the thrust of his argument. 'If you think that all that effort is to be wasted by you firemen throwing the stuff all over the countryside, you're greatly mistaken.' He thumped his, now clenched, fist on the desk. 'Do you understand me young man, it's just not good enough. You will trim your tender, as is your duty, and don't come up with excuses about rough junctions. You know what the road is like and I expect you to have your engine in a safe condition to counter it. Do you understand that?'

I acknowledge what he was saying and apologised for my tardiness in the matter. 'I have your record here and fortunately it seems clean,' he said, having calmed himself. 'You will consider yourself as verbally reprimanded on not having your engine in a safe and fit condition for the road and if you are ever brought to notice on a similar matter, you can expect a lot more than a VR.' He looked me critically with stern eyes, 'Have you got that Fireman.'

'Yes Sir'

'Good, now go away.'

However, just as we were getting to the crunch point, the whole thing seemed to fall apart.

'Were there many casualties,' I enquired meekly.

'What! What are you talking about?'

'Were there many people injured by the coal?' I repeated, feeling sure we had missed the whole point of this interview.

'I know nothing about people injured' he replied. 'The only casualties, as you put it, were railway property. Your little escapade bent some pipe work as I understand and S&T were called out – Now get out of here.'

I did not need a third request to remove myself, and complied with his bidding as fast as I decently could, closing the office door behind me. Just briefly I stopped outside that office, my mind still getting to grips with what I thought should have happened and what actually did.

Those thoughts were soon interrupted by the pretty voice at the desk, 'Oh you're not dead then,' she mockingly enquired with more than a hint of disappointment in her voice.

'No, sorry. No blood today I'm afraid,' I assured her and walked away as briskly as I could. Through the outer office, with everyone busy getting us more coal, apparently, then down the dark stairway three or four treads at a time.

Out once more into the world of hot oil and steam. Of sulphurous smoke and coal dust. Back out to where I belonged. I stopped to speak with no one, but hurried across the coal road and into the ash-laid car park to the side of the shed. There, near the end of a row of cars, mine was parked and into its driving seat I slid. Out of sight of colleagues, or anyone else, I sat in that car to think through what had happened. All that worrying, all that sleeplessness. Those thoughts, certain of manslaughter and criminal injuries charges. All the time nobody was dead, nobody was injured. I didn't know how they could have escaped, but they did.

As I sat there – still a teenager in a man's world, elevated due to circumstances, to the proud position of main-line engineman – my knees trembled, my hands shook and I only just about convinced myself that men do not cry; and all in sheer and utter relief.

7
All in a Day's Work

A view across the works yard from Campbell Road. Several engines are awaiting the call into the works and the Works Shunter is busy about its duty, probably driven by old Tom Featherstone, one of the permanent works drivers who had come off the main line due to health reasons. Conversely, his fireman would probably be a very junior man on the first few rungs of the long promotion ladder.

Looking back on a way of life from a distance of well over fifty years, and from the perspective of a much older age, it is tempting to visualise only the fine and sunny days and conveniently forget the inclement weather that we had to battle at times.

Today's drivers sit in their enclosed, heated cabs, complete with windscreen demisters and ship-to-shore radios together with powerful 'searchlight' colour light signals that can be seen for miles. It is a whole world away from the footplate of a steam engine and semaphore signalling. Then the weather did have a very marked effect on your daily work.

During the mid-1960s there were still a few of the older Southern Railway-designed engines in service that had a lower tender profile, so that footplate crew were to all intents and purposes working out in the open. Behind the cab and above the top of the tender was open space.

The classes of engine that I worked on in that category were N and U (both known to enginemen as 'Woolies'), S15 and I think from memory there were some Qs around. Also, any Western visitors would also come under that description.

Of course, I also have happy memories of cantering through the forest on a lovely early summer's morning with a Wooly startling the grazing deer and ponies as we swept around the curves at Wood Fidley. But sunny summer mornings did not come every day: rain, frost, snow and fog were also part of the norm.

Rain was not so bad if you were running fast and had a storm sheet rigged; but if you were travelling slower than the rain itself when working a heavy slow freight and you discovered that you hadn't actually got a storm sheet to put up, then the rain and hail could drive right into the cab and there was not a lot you could do about it. Your back was cold and often wet, while your front was roasting hot as you plied your shovel. Putting on the injector to replenish the boiler required you to look over the cab side to check it was picking up correctly, which again was an opportunity for a face wash.

A study of rebuilt Bulleid No. 34037, formerly *Clovelly*.

Life was, of course, a little more comfortable with the more modern engine types; the Standards and Bulleid classes all had high-fronted tenders and most had a rubberised canvas flexible connector across the top of tender and engine. This afforded the crew as much protection against the weather as if on a tank engine. However, there were still plenty of times when you had to lean out of the cab into the full fury of the English weather. Then, of course, there was the fog.

Because of the amount of heavy industry belching out steam and smoke into the atmosphere, fog – especially in the London area – mixing with manmade pollutants became smog. At times it was a dense, impenetrable grey veil that blocked out the world, so that signals lit by a single paraffin lamp were unseen until you were actually next to them. These were very difficult and dangerous working conditions, and, as you were in a steam engine also giving off steam and smoke, it was actively adding to the mix by its very being. At these times the order went out to commence 'double block' working. As the name implies, this increased the safety margin between trains from the normal one clear block section between any successive train to two. In other words, it effectively halved the number of train pathways available. The timetable also became 'get there as soon as we can'.

As the Southern was very much a passenger-orientated railway, its freight operations were kept separated from the passenger side as much as possible. So in the very busy commuter periods, morning and evening, the last thing you needed was for there to be a lot of freight workings still occupying much-needed pathways.

As mentioned, although the local pick-up-type freights operated during the day bringing and delivering local area freight to and from the main marshalling yards at Eastleigh, the heavy long-distance traffic between the main marshalling yards were always run overnight. The theory was that by the time the return freight from the distant yard was reaching its destination, the morning commuter traffic would just be starting and the traffic paths cleared for them. However, the dreaded fog always had a major impact on normal working practices.

Fog always meant you were late arriving at your destination. By the time you had completed the loco requirements, hopefully had a meal break and coupled up to your return train, you were very late leaving with the return working.

No. 34034, formerly *Honiton*, powering past the shed at Eastleigh with a Waterloo train. This engine is in the disgraceful state of cleanliness that was so typical of the last couple of years when there were no more cleaners employed. Whatever the state of the exterior – and this is actually a green engine – you may be sure that the inside of the cab is clean, the backhead shining black with copper and brass fittings clean and bright.

At the time the rest of the world was out of bed and needing to get to work, the freights were still a long way from home and suddenly very much in the way of things. It was no surprise to any of the crews working slower heavy freights that in these circumstances they would find themselves sidelined. In fact they would expect to be put into a yard somewhere, or held in a refuge until a clear path was available. It was not a question of 'if' but one of 'how far will we get before we are shut in somewhere'.

For the enginemen this presented the obvious problem; having worked hard all night, they would very much like to just get home, and now they had no idea how long they would be sitting waiting for a pathway.

For the fireman another set of problems was that his fire would be heavy ready to supply the demand for steam that a heavy train uses. But now suddenly the boiler demand was zero, but the fire was still built up. Under these circumstances all you could do was come to a stand with fire doors wide open; the dampers tightly shut and injectors turned on to put as much water into the boiler as the level would allow, all in the hope of keeping the engine quiet and not wasting too much hard-earned steam.

Usually there was the need for the fireman to walk to the signal box under rule 55 to report your presence to the signalman and sign the register. In answer to the inevitable query, the signalman would announce that, 'We've got the block on mate.' Meaning that double block working was in operation and freight was in the way. The answer to any question as to how long the freight was to be held was usually, 'Dunno mate, could be hours.' So there you were and there was absolutely nothing you could do about it.

You had to stay with your engine; it in was in fact a criminal offence to leave a pressurised vessel – i.e. a steam engine boiler – unattended. Remember too this was in the days well before the advent of mobile phones, in fact most people didn't have any a telephone of any sort, so there was no way of letting anyone know you were going to be very late home. But such unavoidable delays were taken as just part of the job and you just had to accept it as a fact of your life. Conversely, the inevitable overtime pay was always useful.

A large shed, like Eastleigh, employed an equally large number of staff from both mechanical and labouring shed-bound employees to the various grades of footplate crews.

Quite naturally too there were very many differing characters among those employees, with a number of very senior drivers schooled in the LSWR mentality whereby a driver of their seniority was in fact infallible, untouchable and in total command of both his locomotive and his fireman, whom he could treat as he felt fit. (Fortunately these men were few by the early to mid-1960s.) At the other end of the spectrum were the younger drivers who were keen to show that they could still wield a shovel and give their younger firemen a chance to do the driving under their strict supervision.

Whatever the characteristics of the driver and fireman, they were put together as a permanent team and had to work out their differences. Just occasionally, a driver really could not get on with his fireman and would demand a change. I am not aware that a fireman had the same recourse.

Generally speaking, pairings would quickly settle down as each man learned the capabilities of his mate. The fireman had to provide steam power to the demand of the driver's style of driving. He therefore had to have a good understanding of how his driver would handle any particular type of train, as well as how to manage the type of engine they were working and, of course, knowledge of the route. These two men worked together in the hot confines of an engine cab day in and day out all the time that both were in the same link. Quite often these pairings extended into social contact outside of work and on occasions engendered friendships that became lifelong.

When one of the men moved up a link, then the pairing would be broken and new ones made. It was usually the fireman that moved up. Once a fireman had a driver's post, his progression would be fairly slow and depended mainly upon more senior men retiring from the top links. On the other hand, firemen had a whole variety of reasons for leaving. For a start, the work was hot, hard and the hours very unsocial. It was definitely not conducive to a young man's desire to woo the ladies. Indeed, many young men left the service before getting anywhere near a driving job and others moved within the railway to a less-demanding job. Because there were also regular vacancy lists posted, the more senior firemen could even consider moving to another part of the system in order to get a driver's post quicker. All in all, a fireman's chances of promotion were considerably better than a driver's, and thus it would more often be the fireman moving up and leaving his driver behind. This happened to me in every case.

I was also very lucky in that I managed to get on very well with all my drivers, even though the first one, in the Ashpan gang, was miserable old Jimmy Thomas – he who hated cleaner boys, and indeed most other people. He was also the butt of various pranks played upon him by disgruntled persons from his days on the main line.

Jimmy would come to work each day on his old push bike, which he parked against the tall yard lamp post just outside the Ashpan Gang's cabin adjacent to the coal stage. Why he never parked it in the bike shed like everyone else, I do not know. But early one morning – about 5am when he decided that he could quietly disappear and leave his fireman (me) to finish the few remaining jobs before relief at 6am – there was a major commotion because old Jim could not find his bicycle. It was no longer in its usual spot. Apparently someone, whose parentage was in dispute, had made off with this most-prized possession.

After a while it was spotted, still against the same lamp post, only instead now of being at ground level it had been hoisted to the very top of the lamp and lashed securely in place. Who the culprit was, I know not. Neither did anyone ever admit to seeing a person climb the lamp post ladder with a bicycle across his shoulders.

From the Ashpan Gang I moved away from the limitations of the engine shed and into the confines of the Works shunters and Carriage Works shunter; carriage siding and station pilot, and Pre-Assembly Depot shunter. As the name implies, this later establishment made up track and pointwork in pre-assembled sections ready for installation into the network; it also went by the name of the 'Madhouse' and consequently the engine that worked here was known as the Madhouse Shunter.

The two Loco Works shunters employed, at that time, were USA-class tank engines, known to enginemen as 'Yankie Tanks', as was the Carriage Works Shunter. These duties were the domain of senior drivers who, like the Ashpan Gang, had elected to spend their final years before retirement on less-stressful duties due to poor health. The station pilot and Madhouse usually had a very junior driver just starting on his promotion ladder as a driver. All the other shunting duties around Eastleigh were single-manned diesel shunters except, as related elsewhere, when it was necessary to temporarily substitute the usual diesel for a steam shunter that required a fireman.

I cannot now remember what the rostering arrangements were for these duties as I certainly never had a regular driver on these turns, which you would do for a week, then go on to a different duty with a different driver.

During the early 1960s, many of the local passenger duties were taken over by the Hampshire Diesel Electric Multiple Units, known later under TOPS system as class 205, but originally just called 'De Mu' by the drivers. These units had caused the withdrawal of most of the class M7 tanks and all the T9s, and also took a lot of work from firemen as they were single-manned units. They also initiated a revision of the work structures for footplate grades as a new link of duties was created covering all the various routes now operated by single-manned DEMUs.

The remaining steam work was thus reorganised into five links of men, not including the dedicated shunters and shed enginemen. They were the Tank Gang; Bournemouth Link: Holiday Relief Link; Spare Gang and Main-Line Link.

The Tank Gang worked mainly – as the name suggests – on tank engine turns, including all the local 'pick-up' freights and local parcels traffic. Their area of operation was Winchester, Romsey, Fareham and Brockenhurst,

The Bournemouth Link worked pretty much everything, as the name implies to Bournemouth, of course, as well as certain turns to Salisbury, Basingstoke and Portsmouth.

The Holiday Relief Link was designed to cover men on annual leave plus a certain number of duties allocated to the link, including, I think, the Chi. (Chichester) Goods. This meant that, although the drivers and firemen were paired, they only actually worked together during the regular booked duties and with other partners when covering for holidays.

The Spare Gang was mainly always covering for sickness and holidays, additional work and anything that could not be covered by regular duties.

Finally, the Main-Line Link was for the most senior of both drivers and firemen and would have the bulk of the London and Weymouth work.

A study of an original Bulleid Pacific on shed at Eastleigh.

A surfeit of power for a pick-up freight, Urie S15 No. 30496.

No. 34042 *Dorchester* ready to go off shed.

In practice, the acute shortage of able firemen during the final years of steam meant that above the Bournemouth Link, a fireman was basically allocated wherever he was needed on a weekly basis. Although my driver was a gentleman by the name of 'Jim' Gates – whose given Christian name was Edward – I rarely worked with him. Along with all my fellow firemen from the Holiday Relief Link onwards, we were all essentially in the Spare Gang. As such we all worked regularly to Waterloo or Weymouth and to Feltham, which a generation earlier would have been unheard of until you were a senior fireman.

The system of permanently pairing a driver and fireman was a very good one as previously described, but just occasionally it did throw up some interesting combinations, especially as firemen were promoted younger due to the critical shortage of that grade.

Hence middle-aged men found themselves paired with teenage firemen, which was not usually a problem as in my experience the drivers were only too keen to show their youthful workmate how to get the very best out of themselves and their machines, together with the all-important brotherhood of that elite band of railwaymen that were footplate crew.

While you can teach the boy how to do a man's work, not all youthful male hormones can easily be suppressed and young men will always be young men. One such pairing was with a real gentleman of a driver who was also a most considerate driver to his fireman. He was also a Christian who was never heard to utter profanities or indeed any sort of bad language. His own regular fireman in the Bournemouth Link was a young man, not a lot older than myself but who was a bit of a lad with an eye for the young ladies – which, of course, is something of which a lot of us were guilty.

On one particular day they drew into Southampton Central on an up London semi-fast and would be due to be relieved at the Eastleigh stop. They were also booked to take water at Central, which was common practice. As was also usual, the engine stopped in such a position that the fireman was required to hold the bag (the leather water pipe) into the filler hole as it was not directly in line to be able to hang down without assistance. The driver as usual was on the platform manning the water valve and waiting for his mate to shout out when the tender was full.

All went well until a party of pretty young ladies in the short skirts of the time and, consequently long legs on display, walked up towards the engine looking for an empty compartment to sit in. This now presented a major dilemma to the young man on the top of the tender; simply put where did he look next? Hormones won and be became distracted from his duty holding the water pipe in the tender. In his quest to get a better look at the scenery on the platform, he altered his stance to the extent that he allowed the water pipe to become kinked on the edge of the filler hole and the whole pipe then reared up under the pressure of the now restricted water flow and emerged completely out of the tender tank, cascading cold water over the side of the tender and on to the unsuspecting driver stood nonchalantly at his post below. Suffice to say that gentlemanly behaviour and religious norms were temporarily suspended, resulting in both profanities of the most vitriolic kind being directed at the hapless fireman atop his tender. The reaction of the females was not reported – it was probably best not to ask.

As for my own experience with regular drivers, this began with driver Peter Bevan in the Tank Gang, indeed I consider myself very lucky to have had some really good pairings in many ways. I think I was still only sixteen when I joined up with Peter and he too was a young driver in his early thirties. He was a very laid back and easy going sort of individual who was rarely phased by anything. It was a joy to work with him and have him teach me my craft. We did all the local pick-up freights together and the local van trains, and generally had a wonderful, carefree time. My abiding memory of those days is working the early morning pick-up out to Totton via Northam yard and back via Redbridge sleeper works and Millbrook yard. This turn, 'the Dodger', set off from Eastleigh yard at about 5.30 in the morning. I can still see us now, waddling down the main line on a Midland Tank (41xxx) with about thirty wagons in tow. Rattling our way through an otherwise silent and deserted St Denys station with Peter sitting on his side and singing 'Danny Boy' at the top of his voice. It is the little things like that that just stay in the memory. As I wrote these notes in 2021, Peter and I still met up socially when the circumstances allowed and even though he was in his late eighties and his former fireman in his early seventies, he still referred to me as 'my little mate'. Sadly Driver Peter Bevan finally got his promotion to the Greater Celestial Engine Shed in December 2021.

All too soon it was time for me to move on up into the Bournemouth link, which actually did the bulk of the steam work at Eastleigh. Apart from working everything from freight to stopping passenger and semi-fast passenger to the seaside, we also did regular duties to Portsmouth, Salisbury and up the main line as far as Basingstoke.

Of course, a new link meant a new driver to work with, but I didn't know who this person was. I only knew that from a certain date I would be expected to move up and take my place in the Bournemouth Link.

As it happened, I was preparing our engine one day in the week before I was to move and my driver, Peter, joined me on the footplate and announced that he had just had a chat with my new mate. Apparently driver Dave Batten had sought out driver Peter Bevan to enquire what sort of fireman he was about to be paired with.

At that time I had no idea who driver Dave Batten was and was intrigued and dismayed to learn that he was called 'Thrasher Batten' due to his driving style. He was known as a fireman breaker who seemed to take great pleasure in making the toughest of firemen quake in their boots at the very mention of his name. That's what I was told, and that is what I looked forward to being paired with the following week.

Having got over my disappointment about my new driver's reputation, I made the conscious decision that I was not going to be defeated before we even started and I would do my very best to be good enough.

I have already recounted my first day's work with Dave Batten and that pretty much set the tone of our work together. His style was physically demanding and you always had to think one step ahead. We had some pretty close shaves almost running out of steam with indifferent steaming engines, but only once did I have to ask for a few minutes extra time in a station to get the boiler back into order before we attacked Sway bank.

No. 73110 *The Red Knight* powering along the up main with a service from Bournemouth, approaching Winchfield.

34073

Happier times at Eastleigh. A pair of Maunsell Qs at the rear of the shed: No. 30532 modified with a multiple jet blast pipe and wide chimney and No. 30530 with the original single chimney. Not one of Mr Maunsell's best designs, in fact, in the words of their drivers 'they couldn't pull the skin off a rice pudding'.

It happened that on this particular day we were working a stopping passenger from Southampton Terminus; all stations to Bournemouth with an N-class engine. Normally these engines steamed easily; basically you just dumped the coal under the fire door and the shaking of the engine did the rest. But this particular engine just wasn't having it. There was clearly something wrong with the draughting and I was forced to use the fire irons continuously. This, however, made no impression on driver Batten, who continued his normal driving style to the extent that by the time he shut off steam for the Brockenhurst stop, the water level had disappeared out of sight at the bottom of the glass, and as the brakes were applied fully there was a very real danger that the forward surge of water in the boiler would uncover the firebox crown and blow the lead plugs. As we ran in, I had both injectors on full with the fire doors shut tight in case there was a loud bang and about 80lb pressure of steam.

We came to our usual shuddering halt with my driver leaning out of his side of the cab, completely oblivious of the state we were in. We were, of course, early and I interrupted his thoughts and brought his attention to the state of the engine and told him I needed some time to get steam back and water in the boiler. Far from being in any way upset that I had my engine in such a state, he was instead overjoyed, even to the extent of calling out to the guard that he had beaten his mate and we had to have a 'blow up'. He had won this particular encounter. Stopping passenger trains are very difficult work if you have an engine that is not steaming because the constant high demand for steam at the restart after each stop and usually the two minutes at the station is insufficient to get much recovery.

That was the only time I had to admit defeat, although on one other occasion I came pretty close with a 76xxx standard that just would not respond to anything I tried and I had a torrid trip to Bournemouth.

One the latter occasion we were booked to clean the fire and have the same engine for the return working, and it was then that I quickly discovered the reason she was not steaming. A completely blocked tube plate and a smokebox – supposedly

Opposite top: **An interesting picture showing a period when the 'big engine side' of the coal stage had gone beyond use and all incoming engines were sent across the turntable to the 'tank engine side' for fire cleaning and water. Coaling was done on a temporary stage using the diesel grab at the rear of the shed. Also evident is the new concrete and wire fence erected between the ashpit bypass road and the car park. This was all part of the rearrangements in conjunction with the extensions to the diesel depot. The red brick building, which at this time was probably out of use, was the firing school classroom. Behind is the bike shed.**

Bottom: **Every picture tells a tale! No. 34073 *249 Squadron* has stopped just right for water so that the fireman atop the tender only needs to allow the pipe to rest against his body to make sure it does not kink on the side of the filler. However, is the fireman watching what is happening with the water situation? No – he appears to be more interested in what, or who, is happening on the platform! Notice also the absolute rubbish he has to work with in the tender. This was so typical of the last years of steam, and yet we were still expected to run a demanding timetable.**

self-cleaning – but with ash almost up to the chimney. So while my driver was having his tea and sandwiches in the drivers' cabin at Bournemouth Loco, I was cleaning the fire and ashpan, cleaning out the smokebox and managing to get assistance from the fitters and boilersmiths to have the tubes rodded clean and the self-cleaning grid realigned. I then had about ten minutes to have a cup of tea and my sandwiches before we were off again.

Knowing what I had been through, my driver very kindly, offered to let me drive back home and he would show me how a locomotive should be fired! My natural instincts were to remind him that as I had just spent my entire meal break putting the engine back into good order, she ought to be steaming like a kettle and there was no need to teach me how to do my job. However, the chance to do the driving back home outweighed the trauma of having the driver think he was better than me at making the engine steam. It also gave me the opportunity to give him a taste of his own medicine and I openly admit that I thrashed the engine all the way home and no matter how hard I drove her, the steam gauge never wavered much from 20psi below the blood line, with driver Batten sat on the seat grinning from ear to ear.

Another little quirk driver Batten had was sounding the whistle every time we were getting to the top of Sway bank and into the cutting before the station. He latter informed me that his sister lived in one of the houses that backed onto the line at that point and it was just a greeting between siblings.

Away from the work environment, I shared some social time with his family as he was a keen fisherman and owned a small rowing boat that was kept at Netley foreshore. Consequently, I spent a couple of Sunday mornings fishing with his young son out on Southampton Water. I regret that it was not a sport that I particularly took to. I just did not like the idea of threading a poor little worm on to a hook, so the son did the deed for me; and when you did catch something and it was wriggling on the end of you hook – what were you supposed to do with it?

I would sum up my experience as Driver David Batten's fireman as having experienced the 'worst of drivers, but the best of mates'. Eventually though the time came for me to be off again, up into the Holiday Relief Link and a whole new set of rosters and, of course, a new driver.

Nominally I was paired with Driver Edward Gates – known to one and all as Jim Gates. Not that we did much work together as by 1966 and 1967, once above the Bournemouth Link, firemen were just allocated any duties that needed covering, so we were all effectively in the Spare Gang.

Sway in the New Forest. Not all the forest is covered in trees and vast tracks are in fact open heath land as here, looking towards Lymington Junction. No. 76031 is powering up the bank and entering the cutting just before Sway station with a Southampton Terminus to Bournemouth stopper.

8
Last Rites

Hindsight is a wonderful thing and it is easy at this distance for critics to say the Modernisation Plan of 1955 was ill-conceived. Naturally, the armchair or line-side observer of the steam locomotive mourned the passing of that form of traction. It is also true to say that those of us who were trained as steam locomotive enginemen also regretted the demise of our beloved steam engines, although in such a macho world as the footplate such sentiments were never openly expressed. We talked gaily of being able to work trains in the future with little effort and no dirt or excess heat; of being in a nice warm cab rather than lashed by the elements. Life was going to be grand. Little did we really understand the full implications or appreciate the dire commercial reasons driving that change.

The inter-war years had seen a slow but steady leaking of trade to road-borne traffic. The war itself had seen the railways take a heavy burden of military traffic and sustain damage by enemy action. Maintenance was not kept up to standard due to lack of manpower, such that by the cessation of hostilities the railways were generally in a poor state.

Much post-war time and financial investment brought some semblance of the former prestige back to the railway, but also saw the increasing threat from private car ownership and flourishing road haulage companies who could deliver door to door without transhipment from rail to road. Inland air travel also started to become a threat to long-distance rail travel for business classes.

In addition, the price of coal was rising in line with the greatly increased production costs. Hitherto it had been the availability of coal at viable costs that had made the steam engine such an attractive method of propulsion, despite in reality being woefully inefficient.

The abundance of labour in the pre-war years now also gave way to a shortage, due in part to war casualties and a preference for young men, in this brave new world, to shy away from hard and skilled work in favour of less-demanding, usually office-bound, employment that had far more social hours.

End of the road for U class No. 31803 with its 4,000-gallon tender. Front of Eastleigh shed.

'Yankie Tank' No. 30069 in a very dilapidated state besides a run-down 'big engine side' of the coaling stage. Even the water column is out of use and broken.

Therefore, with falling revenue and a dwindling skilled workforce, the nationalised railway had no choice but to make sweeping changes to counter the millions of pounds' deficit it incurred.

Experiments, both pre- and post-war, to evaluate the possibilities of alternative forms of traction had proved both diesel and electric propulsion to be viable. Diesel engines are expensive in their initial build, but running costs and servicing requirements are attractive provided that ample supplies of diesel fuel can be guaranteed. Similarly, electric traction was efficient and very clean, while servicing requirements and daily maintenance were minimal.

The alternative, good old-fashioned steam, was comparatively cheap to build, as construction would be in-house using the railways' own works and skilled expertise with over 120 years of experience.

Diesel fuel also has to be imported, while electric relies on a costly infrastructure to support it. Coal, though, was still available, albeit at increasing cost.

The initial modernisation plan envisaged a programme of new-build steam to replace life-expired units, while at the same time building limited numbers of diesel engines for evaluation before deciding which units to proliferate as a replacement for steam.

The exception would be on the Southern Region, which under the ownership of the Southern Railway and LSWR before had long since opted for electric propulsion on its intense inner suburban commuter network. Unlike other regions, the Southern was predominantly a passenger carrier with freight taking a secondary role. Freight workings tended to be limited in number but loaded to maximum capacity.

It was therefore only an extension of existing policy to advocate electrifying the old LSWR main lines to Southampton and Bournemouth, using the third rail system.

Extensive dieselisation of the Southern Region was never part of the modernisation and those diesel locomotives that were allocated to the region were intended primarily for freight and operation over those few lines that were not to come under the first stage of electrification modernisation. The subsequent political intervention that saw wholesale withdrawal of perfectly serviceable steam locomotives in favour of untried, often unreliable or inadequate diesel traction, did not greatly affect the Southern.

The subsequent programme of works associated with electrification was complex. It could also not happen quickly due to the sheer size of the project, which required electric plant installation and track modifications, the laying of the third rail and an associated scheme to replace semaphore signalling with four-aspect colour light. Signalling power boxes were to be built at Surbiton, Basingstoke and Eastleigh, which would in turn supersede dozens of manual signal boxes along the route. Not a single electric train could operate until the infrastructure was in place and operable.

No. 34089 on shed at Eastleigh. I think this must be before her last major overhaul in the works. Alongside are two withdrawn locos waiting for disposal with their rods off. The gadget alongside is a local product to assist in moving heavy, long objects such as coupling rods.

Therefore the main line from London to Bournemouth and Weymouth would remain totally steam operated until right up to the day of changeover to the new service that inaugurated the electric railway.

Ironically, the Southern, which was always the most dynamic and modern railway due to its predominantly passenger orientation, was also destined to be the last steam-hauled main line in the country by some five years.

While the installation of line-side electrical equipment and colour light signalling did not in themselves adversely affect the normal running of the railway, upgrading the track to replace the remaining bullhead rail with heavier flat bottom rail certainly did.

Wherever possible, trains were diverted while this work took place. Where diversion was not an option, single-line working would be instigated or the section would be closed and the service provided using buses. This latter option was usual after the last train on Saturday evenings and throughout the following Sunday.

For footplate crews this meant a lot of extra work on ballast trains and usually entailed signing on for duty late evening on the Saturday; preparing the engine before collecting a fairly heavy train load of ballast, rails and machinery. The usual procedure was to take the train into the nearest holding point before the site of the work and await the passage of the last service train. Once all traffic had cleared, the engineers' train would be taken into the section that was due to be worked on and then that whole section closed to traffic and handed over to the control of the engineer-in-charge. Very often, with lengthy work programmes, there would in fact be two engineers' trains working, both under the direction of the engineer.

Once the train was on site and all the various pieces of machinery dropped off and wagons positioned as required, the work for the engine crews was easy. Wagons would be required to be moved about as work progressed, but there were also long pauses between movements while the track gangs went about their business with a practised expertise. There was no stopping except for a short meal break and it has to be said that the men in the track gangs worked very hard, most of the task being manual labour.

Up in the engine cab, the hardest job fell to the fireman, who had to keep his engine quiet, but not let the fire get too low, and also have enough head of steam ready to move the train around at a moment's notice.

The final method of coaling with a diesel grab on a sleeper-built base at the rear of the shed. This was in the final few months or even weeks when the 'big engine side' of the coal stage and ashpit road had fallen into complete disrepair. Only the 'tank engine side' of the coaling stage was used and then only for fire cleaning, with the actual coaling done on this temporary set-up.

A foreigner on shed. No. 76033 was originally an Eastern Region allocation. Notice the cab recess for a single-line tablet catcher.

Another view of the last days at Eastleigh with the revised coaling arrangements. Engine No. 34021, formerly *Dartmoor*.

During the winter the big inconvenience for both footplate men was the weather. A steam engine is pretty exposed to the elements when it is stationary, and with the fire left to die down keeping warm was a major problem. Hence the experienced men on these jobs would come equipped with extra jumpers and the heavy serge overcoat that was standard railway issue.

Sometimes due to the length of time and the distance covered in these operations, it was expedient to relieve the engine crews. It was not possible to change the engine as the section was taken out of service, so the only alternative was to change the crew on site. The new crew would travel to site in the old Bedford Commer van that was allocated to Eastleigh locomotive shed and the relieved crew taken back to Eastleigh in the same fashion. Where the work site was in the heart of the New Forest, this entailed an interesting and at times hair-raising jaunt down forest tracks to get to the nearest point to the work site.

Weekend after weekend, section by section, the work continued until the entire route from Totton to Bournemouth had been re-laid with continuous welded rail with installations and equipment made ready for installing the electric third rail. New colour light signal posts were erected and a completely new power signal box built at Eastleigh.

Having completed the heavy work of track relaying and storing the third rail with its associated insulation pots by the track, the actual electrification work could be undertaken by specialist gangs working without disrupting normal service.

It became increasingly clear that not only were the days of steam traction numbered, but the whole face of the railway that we knew would be swept away forever. The growing number of diesel locomotives, the closure of less-profitable routes and general track rationalisation allied to the wholesale closure of numerous signal boxes along the main lines all spelt out massive changes in the railway environment and working practices.

It was in this era of change and uncertainty that the engine crews were determined their beloved steam engines would bow out in a blaze of glory.

If this was to be the final chapter in Southern Steam then it should really be written in bold print, for despite the decrepit state of most of the steam fleet due to lack of maintenance and capital expenditure, the engines were pushed to their limits wherever possible and rarely could any loss of time be booked to the engine.

As an example, No. 34089 *602 Squadron* of the Battle of Britain class was always a good engine and it also happened that she was one of the very last to receive a general overhaul at Eastleigh Works. She was clearly given special treatment as befits her place in the history of that institution and as such she was second to none; she could almost literally fly! Several other engines were also in good form.

I still have the small notebook that I habitually carried during my normal duties. Within is a note of a particularly fast run to Waterloo from Southampton, which I scribbled down as it took place.

This happened on 13 October 1966 with a particular mate who was not my regular driver. We were booked to work an up fast from Southampton. The first part of the duty was easy; a comfy ride 'on the cushions' from Eastleigh to Southampton Central as passengers to await the arrival of our train, which we were informed was running late due to engineering works at Bournemouth.

Following a hard sprint up from the seaside, our train arrived with No. 34018 *Axminster* at the front. During the usual task of taking water and trimming the coal, it was established that she was in fine fettle and steaming well. My driver then announced his intention of arriving at Waterloo right time.

Armed with this information, I scribbled down the names of most of the stations and timing points en route in my notebook, with rough columns for speed; steam pressure and approximate regulator position. Although it was apparent that I was about to be very busy that day, these are indicators that would be very clear to me from the fireman's perspective. I would have no easy sight of the exact position of the regulator, the position of the cut-off or indeed the cylinder pressure, and certainly no time to note them.

We pulled out of Southampton Central with eleven coaches on at 6.20pm – sixteen minutes late – and proceeded to run as hard as we could right from the guard's whistle.

I give below the log of that run as seen from my side of the engine:

	Speed	Regulator	Boiler
So'ton C	0	–	240psi
Northam Junct	15	shut	250
Swaythling	43	full/2nd 240	
Eastleigh	55	full/2nd	230
Shawford	58	full/2nd	230
Winch	56	full/2nd	220
Wallers Ash	58	3/4/2nd	225
Mich	61	full/2nd	240
Worting Junct	65 speed restriction	shut	250
Basin	79	full/2nd	245
Hook	85	full/2nd	240
Winchfield	97	full/2nd 240	
Fleet	105	full/2nd 240	
Farnborough	90	full/2nd 235	
Brookwood	95	full/2nd 230	
Woking	97	1/2/2nd 230	
West Byfleet	98	full/2nd	220
Weybridge	90	full/2nd	225
Walton	92	3/4/1st	230
Hampton C Junct	85	3/4/1st 225	
Surbiton	86	full/2nd	220
N. Maldon	76 (60mph restriction)	full/1st 225	
Wimbledon	70 (60mph restriction)	full/1st 220	
Clapham Junct	45 (40mph restriction)	shut	200
Queens Road	45 (60mph restriction	1/2/1st	205
Vauxhall	39 (40mph restriction)	shut	180
Waterloo	arrival	shut	160

Booked arrival time was 7.34pm; we arrived at 7.37pm. – 3 minutes late. Had we had a clear road through Queens Road, we might well have arrived right time.

Just two weeks later No. 34089 would give me a furious, adrenaline-flowing run from Basingstoke to Winchester.

Driver John Stratton of Eastleigh, alas now deceased, was not a man who knew any form of fear. During the week of 24-29 October 1966 I was paired with him on a duty that entailed preparing our engine, working a freight to Basingstoke, then relieving the crew on the down Bournemouth paper train. This we were booked to work as far as Eastleigh, with a stop at Winchester.

For reasons that I can no longer recall, our freight to Basingstoke was cancelled every day and we were left with the easy task of taking the engine light to Basingstoke shed, cleaning the fire and leaving her ready for the next part of her duty, which was an early morning commuter train to London.

What should then have been a leisurely stroll to Basingstoke turned out to be a mad dash as driver Stratton seemed to wish to get to that town as fast as our wheels would take us, irrespective of the 60mph maximum speed allowed for a light engine.

The return journey with the down paper train was performed with the same urgency each day. Clearly Driver Stratton was having his final fling with steam. The route from Basingstoke all the way down to sea level at Southampton is a steady descent with a very fast section through Winchester Junction, while even there the speed was only checked by the necessity to stop at Winchester City station.

Each day, with very indifferently riding Standard class 5 engines, the speed was urged higher and higher as we tore down through Micheldever, through Wallers Ash to Winchester Junction, where steam was finally and reluctantly shut off and full braking employed in the hope we might just manage to stop at Winchester – a mere 3 miles down the line.

On the Thursday of that week, engine No. 73155 was coaxed up to 95mph before steam was shut off.

The 73 Standard engines are two-cylinder machines not renowned for their smooth riding characteristics, and 73155 was not the best of this class. By the time we were racing towards Wallers Ash it became impossible to stand on the bucking, swaying footplate and it was completely out of the question to balance between engine and tender to fire her.

Friday came and to the sheer joy of my driver, the paper train drew into Basingstoke platform with Battle of Britain class No. 34089 at the head. We both knew from our individual experience that this engine was a flyer and it was no surprise to me when my mate asked that I should give him all I could as he was about to see just what she would do!

We left Basingstoke with the usual nine vehicles behind the tender. As soon as the engine had settled into her stride, she was opened up to full regulator with the cut-off shortened as speed increased. We were travelling at over 70mph before we started our descent towards Winchester. But this was no rough old Standard engine; she was a thoroughbred and rode beautifully. There was no difficulty in keeping pace with her ferocious appetite for fuel as she pounded her way down the hill.

Micheldever Station flashed by at a full 100mph, and still we were accelerating under the incredulous gaze of the signalman. Clearly the inter-box telephone circuit was buzzing as his counterpart at Wallers Ash was also to be seen in the light of his signal box. He seemed to be timing us as we passed, by now well in excess of 100mph.

As we approached the junction, my driver drew my attention to the speedometer. The calibrated indicator was incremented up to 105mph. Then there was a gap at the end of the scale, which probably equated to 5mph, before reaching the stop pin. The speedometer needle was resting firmly against the stop pin!

As steam operation entered its final few months, Eastleigh became a dumping ground for withdrawn engines. Here we see No. 73085, formerly *Melisande*, withdrawn and with her motion taken down ready for towing away.

A rail enthusiast special over the Alton Road with No. 92220 *Evening Star* taking the single-track section to Winchester Junction. They seem to have the company of a locomotive inspector complete with regulation trilby hat and gabardine raincoat.

How far past that stop pin the needle would have registered if it was free, we shall never know. As with previous days, steam was shut off and full braking applied as we passed the signalman in the junction box; clearly to be seen with a watch in his hand.

My immediate reaction was to put on both injectors fully to put as much water as possible into the boiler against the forward surge as the brakes screamed in protest against the engine wheels. The lights of Winchester station approached far too rapidly and by the light of the, now open, fire doors I witnessed the spectacle of my driver leaning on the brake handle, willing the brakes to bite harder. This man of solid iron who knew absolutely no fear; this man who had previously spent a large part of the Second World War behind enemy lines fighting in what was to be the forerunner of the SAS, was looking just a little concerned that tonight his train would not actually be stopping at Winchester!

On the platform, lines of men waited with their trolleys to unload the papers at the usual, time-honoured appointed place for the train to glide serenely to a halt and were somewhat disappointed to see us go right past them and come to a juddering halt at the far end of the platform. But at least we stopped!

The guard came up to the engine as the disgruntled paper men traipsed up the far end of the platform to unload. He too was not very amused. 'Blimey driver, you trying to give me an 'art attack. Dunno what speed you was doin' but I thought we was going to come orf the road,' he complained in his Cockney accent.

With ambitions satisfied and a box full of fire, the run down to Eastleigh and relief was a comparatively sedate affair. No complaint was ever made, so it must be assumed that the signalmen enjoyed the exploit as much as we did.

The modernisation plan envisaged steam traction to be employed up to the end of the winter timetable in April 1967. This allowed for the employment of steam-heated passenger stock in the absence of electric heating (ETH). However, the budget for steam locomotive maintenance was slashed and the number of steam fitters and other specialist grades reduced as those men were sent away to retrain on the modern forms of traction.

The reality for those of us trying to run a full service with steam power was abysmal. Most of the engines still running were in a critical state. Many would never have been allowed off shed ordinarily and with these tools we were expected to haul heavy loads and run fast trains. With some engines there was so much steam escaping from leaking glands and unions that it was almost impossible to see ahead until speed reached a point where it would carry away the cloud.

The now preserved No. 34027 *Taw Valley* leaves Southampton Central with a clear road.

No. 34087 *145 Squadron* arrives at Southampton Central from Waterloo. Behind the engine is the power station, which was a major backdrop to the station for many years.

From the fireman's perspective, many of the engines were so bad that they just would not steam. Added to the exceptionally poor state of fuel that was being dumped on the remaining steam sheds, it all meant the poor fireman was hard-pressed to offer enough power for his driver to maintain time.

It is probably true to say that in the last two years of steam on the Southern, the firemen of that era worked harder than any of their predecessors ever did. Use of the fire irons while on the run was sometimes unavoidable, but still frowned upon by most drivers. During this period using fire irons to stir the fire was a frequent necessity brought on by the rubbish we had to fire with. In the event, neither the infrastructure nor sufficient of the new stock was ready at the end of the winter timetable in April. Steam was to limp on to a new date in June. Even that date slipped and the remaining steam was finally withdrawn en masse on 9 July 1967.

During the months prior to that date, the new colour light signalling, with its associated power box at Eastleigh, was connected and commissioned. The new power box itself was cut into the system overnight on a weekend. This did not quite go as planned and the area controlled by the new box was closed to all traffic for considerably longer than the expected 6.00am on the Sunday morning. Consequently, light engines returning from their Saturday duty and due to be serviced at Eastleigh shed were temporarily held in the reception road at Bevois Park goods yard.

My driver and I were bringing an engine back from Bournemouth and were expected to be held at some point as it was known that the Eastleigh area would be closed down. We were not therefore surprised to be stopped at Mount Pleasant Crossing Box and informed by the signalman that we were to go into the yard under caution as other engines were already standing on the reception road. We joined a queue of six of all shapes and sizes. Having prepared our engine for an hour's stay until the new box was due to open at 6.00am, we adjourned to the shunters' cabin to make a can of tea. That little hut was heaving with enginemen all wanting to have access to the kettle, some having already been there for over an hour.

The appointed 6.00am time came and went without the slightest sign of us being allowed home. More engines joined the queue and, despite frustrated telephone calls and threats of mass disobedience, there we all sat, helpless. Finally, at around 11.00am word came through that we were to be under way, but we were required to only use one pathway as testing was still in progress. By this time there were thirteen engines all lined up awaiting their turn to go home. So against all known regulations, all thirteen were coupled together and all thirteen went as one up to Eastleigh and on to the shed. I suspect that if the civil engineer had known he had the weight of thirteen engines running over some of the bridges en route, he would have had some pretty strong words to say on the matter!

A commendably clean No. 34087 at Southampton Central. Sadly, the original buildings and the iconic clock tower have all been swept away in preparation for the concrete office block. Notice also the fireman having to hold the water pipe so that it stays in the tank filler. On this tender there are two rectangular fillers either side of the tender, rather than a central one. Hence the engine is not stopped exactly right for the pipe to stay in the filler unaided.

9
A Personal Recollection of the Different Classes of Steam Locomotives as Working Machines

It may be of some interest if I give my impressions and experience of the various different classes of locomotive active during the final years of steam operation. I will only talk about those machines that I personally worked on and thus classes such as LN, V or N15, which I was too junior to actually do any work on, I will omit. I have plenty of feedback from men who did work on these engines in their prime, but that is only hearsay. Also, I will not consider moving an engine within the shed to be classified as 'working' on them. Thus, although we had a 'Terrier' tank – or as the Brighton men called them 'Rooters' as shed engine on a number of occasions, I would not say that classified me as knowledgeable on their abilities.

Therefore, from memory, we are talking about the Bulleid Pacifics of MN and WC/BB classes, Standard classes 73xxx, 75xxx, 76xxx, 80xxx; 92xxx, Midland tanks (41xxx), USA, U, N, S15, Q1 and, of course, some Midland Region engines and Western Region things.

No. 73029, a green Standard 5, which was a former S&D engine. Although she is at the top end of the shed, she appears to be having her fire cleaned away from the pits.

Starting with the most glamorous; the Merchant Navy class, or as they were known to enginemen Packets, after the class leader *Channel Packet*. Here I can only speak of the BR-rebuilt version, which was powerful, heavy and fast. The boiler design was probably the best there has ever been and was designed to burn indifferent and poor-quality fuel. The south of England is not noted for its coal mines and those that do exist do not produce prime steaming coal. But the Bulleid wide firebox would burn most things; the downside was that they required a very heavy fire on the grate. As with all Bulleid wide fireboxes, you had to fire in a horseshoe; down the sides, under the door and most importantly constantly feed the back corners where the main draft on the fire acted. You could quickly get a hole in the fire bed if you were not vigilant. With a Packet, you also needed to keep putting a few down the middle as well. But you had to have a fire that was as deep as possible and we frequently made use of the half door to fire over.

No. 34006, formerly *Bude*, with her extra-long smoke deflectors coming of the 'tank engine side' of the coal stage. Notice also the piles of new sleepers in readiness to make major track alterations so that diesel locos can access the diesel depot and the new fuelling arrangements directly from the shed entrance lines rather than travelling beyond the diesel depot and reversing back in.

Another delightful view of No. 80152 coming on shed with the all-important headcode displayed. The cars in the car park are also indicative of the period with Ford Anglias, Triumph Heralds and various Hillmans in evidence.

I have heard tales about an LNER A4 in the later preservation era coming out of Waterloo on a steam special with about a foot of fire evenly over the box. Try that with a Bulleid and you will be in deep trouble before you get to even got to Vauxhall! But on the upside, a Merchant Navy would pull anything you gave her, provided she kept her feet and didn't go into a slip. Driving any Bulleid Pacific was quite a skill when pulling away from a stand. They were light on their feet and the driver had to constantly watch the cylinder pressure gauge and try to react by shutting the regulator back before she picked up her wheels. So, starting from Waterloo with a greasy rail and a heavy train, the driver would be constantly pumping the regulator until she got into her stride, not helped as in the third rail area it was forbidden to use the sanders to assist adhesion.

A Packet was like a refined and dignified lady. She took her time to get into her stride, but she would get there and stay there, maintaining a fast speed for mile after mile and generally steaming hard all the time. Obviously that last remark is a generalisation as every engine, even within the same class, would be different dependent upon its state of maintenance and a host of other reasons.

Turning now to their smaller cousins, the West Country/ Battle of Britain class; original and rebuilt.

The original 'air-smoothed' version was a free-steaming, easy-riding machine when it was first out of works after overhaul. Basically, it was a driver's dream and a fitter's nightmare. It had a steam reverser and, as everyone knows, a chain-driven valve gear, which made for free running. The snag was that wear and tear stretched the chain and the graduations on the reverser became increasingly inaccurate to the eventual point of being irrelevant. It was known that an engine could be powering along with the reverser showing a position of mid-gear or even reverse. Common practice was for the driver to lock in the gear at what seemed like a good position according to the audible beat from the chimney and then just drive on the regulator all the time. The usual practice for Southern drivers was to drive on full regulator as soon as possible and alter the cut-off.

On the fireman's side too there were several innovations. Both live and exhaust injectors were on the fireman's side, whereas normally only the exhaust injector would be easily accessible to the fireman without disturbing the driver. The injector controls were grouped so that the water and steam feed valves were next to each other, rather than the usual practice of having the controls in different points in the cab. The fireman's tip-up seat was also actually padded just like the driver's. A steam-operated fire door control was originally provided that was triggered by use of a foot treadle. However, in practice this aide became unreliable and the danger was that the fireman's hands could become trapped in the firebox if the doors closed prematurely, which did happen. This mechanism was disengaged later and the butterfly doors were operated manually.

Campbell Road bridge with both shed arrival and departure roads fitted with third rail and replacement sleepers as required. The previous semaphore Eastleigh up home was situated about 50 yards south of its replacement four-aspect colour light signal with feathers for the up local. No. 75077 is going off shed with the fireman checking that the exhaust injector is picking up properly.

A strangely deserted Eastleigh shed from the rear. Even at this late stage the track is kept clean and tidy. The two engines No. 34010 *Sidmouth* and 34102 *Lapford* conceal the shed lavatory block and partially hide the shed enginemen's concrete cabin. Also in view are new track panels stored ready for use in conjunction with alterations to the diesel depot.

For the first time on a Southern engine, a drop grate was fitted, which was an enormous help to the fireman when cleaning the fire. It consisted of a large square central section of the fire grate that could be opened by a lever and bar in front of the firebox. This allowed the centre section to open to approximately 60 degrees and for clinker to be pushed though the resultant hole instead of the usual practice of digging it out manually with a long clinker shovel and depositing same over the side of the cab. The slight downside of this system of drop grate was that, in later years, it was sometimes known for the mechanism to fail en route and deposit the fire into the ashpan. The view from the cab was a little strange to start with as you were looking down the slab side of the casing rather than a boiler.

The rebuilding of the Bulleids resulted in what was virtually a new engine. All the suspect bits were replaced with more traditional mechanics and the redesign had given thought to how the maintenance fitters were going to work on them. The troublesome chain-driven valve gear was replaced with three sets of Walschaerts motion operated by a good screw reverser. Because of the increase in the weight and centrifugal forces of the additional mechanics, the rebuilds were never quite as free running as their sisters in original condition; even so, they were comfortable and smooth-riding engines and popular with their crews. I would say that a rebuilt WC/BB was my personal favourite locomotive to work on.

The downsides were that, as with all the otherwise wonderful Bulleid boilers, they would burn pretty much anything – but an awful lot of it. Simply put, they were coal scoffers. The injector control, while in a great position for the fireman to use – under his tip-up seat, had a habit of jamming and use of the coal pick as extra leverage was common practice.

The best engine was undoubtedly 34089. This was the last to be given a major works overhaul and I do not know if they did an extra special job on her, but she could fly. No. 34098 *Templecombe* was also a very good engine, whereas 34108 *Wincanton* was the very bane of my life. I could never get the thing to steam properly.

The BR Standard classes were represented at Eastleigh by 73xxx, 75xxx, 76xxx, 80xxx with visiting 92xxx. As the name implies, the cab layouts of the Standards were all the same and wherever possible the equipment was standardised. So, for instance, if you needed a replacement gauge glass from the stores, you just asked for a standard one rather than having one of the right dimensions for the particular class of engine.

The Standards were competent engines but with absolutely no frills about them. They were noisy and generally rough riding. All were based on LMS prototypes but without the niceties and ride qualities of their forbears. During the later years, a Black 5 was booked to work the York–Bournemouth from Rugby and Eastleigh men worked her from Eastleigh to Bournemouth. Once you had sorted out which control did which function (a steam engine does not have labels on the controls, you are supposed to know where everything is) they proved to be fine engines. They rode well and without the rattles, bangs and lurches of a Standard 5.

The Standard class 5 and 4 all had Belpaire fireboxes that were straightforward to fire. All had rocking grates and an ashpan opening lever, which allowed for the ashpan to be emptied without having to go down the pit under the engine. They were also equipped with a self-cleaning smokebox, which realistically meant there was less ash deposited in the smokebox, rather than no ash. However, all too often the little 'SC' plate attached to the smokebox door was seen as an excuse for the disposing firemen not to open the door and check the level of ash and char inside. The result was that as the char built up, the lower fire tubes became blocked and steaming problems arose.

In my opinion, the best of the Standard designs was the 80xxx tank, known to all enginemen as an Eighty Thousand Tank. They were a very capable class 4 engine with better riding qualities than the tender engines due to the rear bogie under the cab. The cab itself followed the standard cab layout for controls and was large and comfortable. These engines were very much the staple of the Tank Gang, along with the smaller 41xxx class 2 tanks.

These smaller engines were known as little Midland Tanks due to their design origin being on the LMS. They were basically a more modern replacement for the old LSW M7 tanks with all the mod cons of rocking grate, mechanical lubricators and a large and spacious cab. They were comfortable and cosy engines with a particular mode of travel but they tended to waddle along the line, a bit like a duck! Not that you particularly noticed this from the footplate, and I guess all two-cylinder engines had a tendency to oscillate from side to side, but it always struck me that the little Ivatts were more prone. You noticed it most if you chanced to look down the side of the boiler and fix your attention on the chimney in relation to the scenery out in the rest of the world. The chimney had a pronounced side-to-side movement compared with the static world at large, especially if she was working hard.

I suppose we must also mention the USA-class tanks, known to one and all as 'Yankie Tanks'. They were bought as war surplus for exclusive use in the Southampton dock system. When they were displaced by the new Ruston shunters (later class 07) they tended to migrate to various shunting roles around the local area, most notably the two Loco Works shunters; Carriage Works shunter and the Redbridge Sleeper works shunter. Often one would also be employed as the Loco Shed shunter.

No. 34101 *Hartland* ex-works outside the top of Eastleigh shed and in steam ready for her next duty.

'Yankie Tank' No. 30073 following a works visit and repaint into Southern Railway Maunsell green. Also just in the picture is No. 80146, which just happened to be the very last steam engine I worked on, on the very last day of steam in July 1967. Behind No. 30073 is the special coaling machine designed to assist coaling for Bulleid Pacific high-sided tenders. I do not recall ever seeing it working.

As the name implies, they were a standard American design of shunting engine – or Switcher in American terminology – designed for easy access and pure functionality. So far as aesthetics were concerned they would certainly not win any award. Everything possible seemed to hang off the boiler, while there was a very large cab that seemed to stretch halfway along it. They did have a useful rocking grate with a curved rocker bar that was neatly stored on the cab floor along with all the other detritus such as oil bottles, bucket, brush, headcode discs and general stuff; all this because I do not recall they were provided with any locker facilities. They also had a very short wheelbase for a six-wheeled engine, ideal of course, as their big selling point was for dock work with consequent sharp curves. They were perfectly adequate for what was required of them, but you wouldn't want to go very fast or very far on them as they had a tendency to jump about quite a bit.

Bulleid's Q1 class of freight engine, known as 'Charlies' due to their original SR number under the Bulleid scheme, C1–C40, were quite something else! Aesthetically they were probably the ugliest British design of engine that ever was built but they were wartime utility engines. This meant they needed to be as powerful as possible and have as light an axle load as possible to give the greatest route accessibility. They were therefore devoid of anything not strictly necessary in order to save weight and constructed in the lightest way possible. In short, they were basic.

Starting with the cab, this was a relatively short and a very spartan affair, with what reminded me of a scaffolding pole as a regulator handle. The tender front was at much the same profile as the cab, which made for an enclosed working area. That was good, except that the grate was fixed and a clinker shovel was needed to clean the fire. With a grate nearly 9ft long and a cab length less than that, it made manoeuvring a hot clinker shovel loaded with dirty fire and clinker that had to be ejected out the side of the cab an exercise in extreme dexterity and frustration. Clearly Mr Bulleid had never had to work with his own creations.

The upside was that all Bulleid boilers were marvellous steam producers and the power to weight ratio was quite phenomenal. They could pull a lot of tonnage but stopping was an entirely different proposition; what is good for power–weight statistics is equally disastrous for the all-important brake force. They just did not have the axle weight to give an effective brake. With a fully fitted train this was not so important as the whole train was braked and did not rely upon the engine. With a loose-coupled or partially fitted freight for which they were designed to work it mattered a great deal!

It is my personal opinion based upon experience working with his engines that Mr Bulleid was a brilliant innovative designer but had no real idea how his machines would perform off the drawing board and in the hard everyday life of the railway. Here reliability was key and ease of maintenance essential.

In hindsight, Mr Bulleid was perhaps best suited to his role at King's Cross under Gresley, where his innovation and undoubted genius flare could be harnessed and channelled into useful contributions. Simply put, he needed someone holding his reins. He just could not see the drawbacks of his designs and the problems they caused, either because of the non-availability of material to achieve his goal or the physical capabilities needed to maintain and operate his machines. Unfortunately, as he was the chief mechanical engineer there was no one in a position to point out the flaws to a man who would take no criticism.

An unusually clean Charley, No. 33019 at the rear of number 1 road.

Another stalwart from the Tank Gang days, No. 41294 at the rear of Eastleigh.

A pair of N-class mixed traffic engines. The nearest, No. 31814, is a right-hand drive example with a 3,500-gallon tender, while the furthest engine is a left-hand drive with 4,000-gallon tender.

Moving on, there were still some of the older Southern engines in use during the mid-1960s, mainly the S15 freight engines and the U and N classes of mixed-traffic locomotives, collectively known as 'Woolies'.

The Woolies were general purpose maids of all work, and very much up to the task. The 'U' class with their 6ft driving wheels were a shade faster than the 'N' class with their slightly smaller driving wheels at 5ft 6in. However, the type of work they were employed on hardly made the slightest bit of difference. In my opinion they were a good engine if they were serviced correctly.

Being constructed in an era that did not pander to labour-saving devices, their 8ft sloping grate was fixed, so fire cleaning was with a clinker shovel and over the side of the cab. The ashpan was cleared with an ashpan rake from down in the pit underneath the engine. By opening both front and back dampers fully, the ash could be pushed out of the pan and into the pit. This was, of course, very fine ash like you would get after a bonfire, and it would blow everywhere at the slightest breeze. The trick was to make sure you were upwind of any breeze so that the ash would blow away from you. Having said that, it rarely worked, as being trapped down a pit under a large piece of machinery, you couldn't get out of the way quickly and any breeze had a habit of changing direction and smothering you in fine ash. The smokebox also had to be emptied with a firing shovel. Again, no self-cleaning smokeboxes here. Smokebox char can, in fact, be the worst if the wind is blowing it about. It is gritty and very uncomfortable if it gets down the collar or into boots. In today's Health and Safety world, you would probably have to wear a protective one-piece suit, safety glasses, hard hat and a mask. In those days we didn't.

There was one fireman at Eastleigh – who shall remain nameless, but will certainly be known to his peers – who I am sure was born with a protective bubble around him. He could complete these tasks and still have spotless overalls and shining boots. The rest of us mere mortals would just get dirty.

The Woolies were actually easy engines to fire with a Belpaire box and a steeply sloping grate. Most of the fuel was put under the door or just in front and the motion of the engine and the slope of the grate would work the burning fuel forward to the front of the box. You just needed to keep one eye on the front to make sure there were no bright spots indicating a thin fire bed that needed a few extra rounds. If you ignored what the engine was telling you and allowed a hole to burn in the fire bed, the primary draught up through the dampers would always take a path of least resistance and your little hole quickly became a big hole. The result was that the draught on the rest of the fire decreased and pressure fall away.

That was all very basic firing technique and otherwise you just had to keep up with the rate at which the fire was being burnt by your driver. As these machines were one of the staples for Bournemouth Link work, my particular driver would be hell-bent on trying to empty the tender! That said, I liked the Woolies and thought they were a good engine.

The S15 was the heavy freight engine of the SR, originally designed by Robert Urie in 1920 with massive proportions, generous bearing surfaces and of very robust construction. The initial batch was later followed by a further twenty-five engines during the tenure of Maunsell in the chief mechanical engineer's office. These later engines were just an updated version of Urie's design with improved draughting and, most notable for enginemen, a Maunsell cab that was an improvement on the Urie version. Both versions were heavily built and when coupled to a large bogie tender had plenty of brake force – which was a great comfort on a heavy, partially fitted, freight.

I do not recall specifically working on a Urie version, but had plenty of experience on the Maunsell engines as these were the preferred type on the heavy Feltham or Nine Elms freights. There was no fuss, no nonsense; the engine just got on with the job, dug in and pulled for all she was worth. They were also one of the few engines that could be put into full forward gear; you could open the regulator to the full second valve and pull steadily without the hint of a slip even with the cylinder pressure gauge showing maximum.

For many decades a landmark in Southampton, the clock tower associated with Central Station. No. 76054 is on a very short van train to Portsmouth.

A particularly favourite engine to the author, No. 34022 *Exmoor* was the Bulleid West Country Pacific that I worked on more than others. There was no particular reason for this, it was just the way the engine diagrams coincided with my working roster. Result – many happy hours spent on her footplate.

Southampton Junction where the 'new' line to Dorchester joined the original line to Southampton at the Terminus station. This is Tunnel signal box. The hoarding behind the engine hides bomb damage from the war and the line from here back to the actual tunnel passes behind some pretty run-down properties and tenements in the Northam area of Southampton. Due to the filthy state of the engine it is impossible to identify which class member she is, possibly No. 35028 *Clan Line*. What we can deduce is that the crew are Nine Elms men as witness the fireman's head gear, viz; a knotted handkerchief.

Pure nostalgia. A panoramic view of Eastleigh station from the shed arrival and departure roads featuring U-class Mogul No. 31804. Right-hand drive with a 4,000-gallon tender on a van train to Bournemouth, a favourite class of engine on a favourite type of working.

As I have already described, it was common practice for a driver to just lean his weight against the regulator handle as soon as he got the clearance from his fireman upon leaving Eastleigh northbound with a Feltham or Nine Elms freight. The sparks may fly out the chimney, but she would just pull. It was on those occasions that you really felt that you were a team of three, physically moving all this tonnage together along the main line: driver, fireman and engine.

Probably the only engine better for freight working on the Southern was a Standard 9F, but there were very few of those and I certainly never had one on either a Feltham or Nine Elms. They were almost exclusively kept for the Fawley petrol trains until displaced by the new Crompton diesels.

Now, of course, we must mention the foreign interlopers. The one turn that we regularly had a foreign engine with was on the daily York–Bournemouth, which Bournemouth Link men worked from Eastleigh to Bournemouth. This was hauled by a Banbury engine, usually a Hall class or sometimes a Grange. These, of course, were GWR engines, so what can a decent LSW-trained man say! They were awful. The worst thing for me was the tender shovel plate was actually at floor level and the fire door was the same, hence you were bent much lower than on a proper SR engine to do the same work. I will admit that they were pretty engines when clean, but that didn't help much; they were still GW.

Later, when Banbury closed to steam, the down York had a Rugby engine that worked all the way through and this gave us our first introduction to the Stanier Black 5. Apart from the fact that by the time we relieved the crew at Eastleigh the fire was pretty dirty, they were good engines and I fully understand why they were popular with their crews and why so many were built. The only thing I was no so keen on was that awful Midland hooter; it needed a proper whistle.

As steam works across the country were being run down, Eastleigh found itself in the position of taking on major overhauls of various classes of Midland engines. Engines entering the works did not do so directly, they first entered the running shed for fire dropping and servicing and were then left dead for a couple of days so they were completely cold before being hauled over to the works along with other engines booked for a visit.

So it was that in 1966–67 we had many unusual engines appearing on shed from the Midland Region. These included Black 5s, 48xxx freight engines, the rather ungainly 43xxx as well as the aesthetically more pleasing Stanier moguls.

All these engines returned to the running shed after their works visit and it was usual practice for Eastleigh to book an ex-works engine on a local running in turn before dispatching them back to their home depot. This gave us the experience of working with some of these machines. It was my overall impression that these were good engines, which is not surprising as they were, in many cases, the forerunner of a later BR Standard design – only not so rough riding, with not quite so many rattles and bangs and generally probably better built.

10
Goodbye to Steam

So it was that Saturday, 8 July finally dawned and we all knew that this was to be our last day working on steam.

There were many comments that this day was welcome; that men were truly glad to see the back of steam and the future could only be wonderful. No more hard work, no more dirt and grime; life would be a pleasure henceforth. It is also true that, given the decrepit state of the machines we had had to work with over the previous two years, such comments were easy to make and easy to understand. And yet, 99 per cent of these were also outright lies. Whatever the outward show of bravado, these men were steam enginemen first and foremost. Their training and every instinct was that of a steam engineman; it was their very life.

On that final day I found myself paired with a bull of a driver, a man not given to displaying sentiment. Our task was fairly undemanding and concluded with shunting vans from both the up and down mail trains, which crossed at Southampton. Thereafter we travelled light engine back to Eastleigh, arriving on shed at about 1am on the morning of 9 July. This was not the usual sprint of men going home, rather a more sedate last journey.

On our arrival we were instructed to take the engine round to the top of the shed and clean the fire as she was to have the sad duty of taking some of her dead sisters over to Salisbury en route to the scrap yards later that morning.

We dropped down on to the top of No. 1 road and came to a stand adjacent to the water column. For one final time, I heaved the, now stiff, column arm round towards the engine and put the pipe in to fill the tanks. That task completed and back in the warm cab, I awaited my driver to go so that I had room to manoeuvre the fire irons and clean the fire. For several minutes my driver just sat at his controls deep in his own thoughts. Presently he stood up, retrieved his coat and bag from the locker, then turned once again towards the engine. With a deep sigh he patted the regulator. He was oblivious of me, as I stood on my side of the footplate, when he said quietly and with deep feeling, 'Cheerio old girl,' then he turned away and left me to my work.

No. 34087 in the rain at Eastleigh; shorn of her nameplates, but still very much in use.

With the fire cleaned; fresh coal applied; backhead wiped over with an oily cloth; brass and copperwork buffed up; floor swept and hosed down; she may have looked a wreck from the outside, but inside the cab she would take her final run in good order.

Then it was over. I too sat for a few minutes on my side of the cab in the soft warm and welcoming light of the partly open fire doors. Steam was gently rising from the freshly hosed floor. Outside steam was also rising gently from the various leaking glands.

Tomorrow would be the start of a brave new order. No more battling with the elements, no more hard work, no more dirt and grime, but no more steam. What terribly conflicting emotions.

Then it was my turn to retrieve coat and bag, my turn to leave, and I too found myself not wanting to go. So, I thanked her for all the joy of hard work she and her breed had given me, for turning the boy into a man, and quietly left her.

Walking across the top of the shed towards the enginemen's lobby, I looked back one last time to see my engine for the day, No. 80146; just a silhouette of filthy dirty, unkempt, ill-maintained machinery shrouded in a mist of escaping steam. It was all over.

Just a few examples of those machines that were once the bedrock of our transport system would find their way into museums for future generations to marvel at, together with the characteristic black serge jacket and shiny grease top hat that were the badge of office for generations of those elite of railwaymen who worked with them.

Although the Waterloo–Weymouth section of the Southern Region was the last fully steam operated main line on British railways, steam traction did not disappear in one swoop overnight. It is true that main-line passenger services remained predominantly steam-hauled throughout the winter timetable of 1966–67 due to the coaching stock being steam-heated and available diesel locomotives being equipped for electric heating only. On the freight side, steam steadily gave way to diesel in the final year, aided by the fact that freight traffic was itself in sharp decline.

The governing factor for withdrawal of steam was the decision not to continue expenditure on heavy maintenance. Thus, as locomotives became due for heavy overhaul and boiler change they were withdrawn. Inevitably this gave rise to the unfortunate situation where many good and useful locomotives were dispatched to the scrap lines while inferior machines soldiered on to the end.

By July 1967 there were only seventy-two steam locomotives operating on the Southern. This last remnant was all withdrawn on 9 July 1967. In order to alleviate the growing

No. 3101 *Hartland* as it heads away from Southampton Central towards the tunnel under the city centre. The driver seems to have left the cylinder drain cocks open for an unusually long time.

Another down Bournemouth working, No. 34001 *Exeter*, powering past Worting Junction with Battledown flyover in the background.

shortage of locomotives for the passenger service, four Brush type 4 diesel locomotives, equipped with steam-heating boilers were loaned to the Southern Region. On these machines it was the fireman's job to operate the steam-heat boiler; a duty that entailed working within the engine compartment while the engine was working flat out and not an ear defender, hard hat or hi-vis jacket in sight.

Despite the appalling state of those steam engines that remained active, they did all that was asked of them and their crews regularly gave 100 per cent effort rather than opt for the easier choice of refusing to work on such machines.

The morning of Monday, 10 July was the start of the new timetable and I was again on an early turn of duty. Ironically, I had just finished a sixteen-week spell without a full night in bed, only to find that under the new roster I was about to embark on a further fourteen weeks of similar duties.

Around 7am that day I had control of a small Drury shunter in the carriage sidings at Eastleigh station, my driver having disappeared somewhere for a smoke and a read of his newspaper. A down semi-fast train pulled out of the station with an electro-diesel locomotive on the front as I was stationary waiting to push some stock into the platform. It was a clear and sunny morning as I idly watched the departing train with a London crew on board.

Previously, with a steam engine, you could open up and pull purposefully out of the platform, not so with a relatively short-length electric locomotive. As the down relief line joined the down main line, there was a gap in the third rail that was greater than the length of such a locomotive. When the engine lost contact with the third, conductor rail, the driver was supposed to shut back the power demand until safely over the gap, when he could put on maximum power again. Clearly this driver forgot.

The peace of my little world enclosed in the warm cab of my engine was rudely shattered by an almighty blue flash as the electric engine left the third rail drawing full current, which did its best to maintain contact through the atmosphere. Immediately following there was a loud bang and a cloud of brown smoke engulfed the engine as the front pick-up shoes gained contact with the next bit of third rail, still trying to pull maximum current. The whole train glided to a halt, straddling the down relief and main lines. Not only had the driver succeeded in blowing the pick-up shoes clean off the engine, but he had also blown out all the circuit breakers on the down line. Thus nothing powered by electric was about to move for some while.

I was shortly dispatched with my little shunter to come on to the back of the stranded train and pull it back into the platform, from where the scorched and damaged engine was removed and sent to the shed for attention.

This early incident brought home the need to learn a whole new set of rules and track knowledge. You had to know where the gaps were and their effect in relation to the machine you were driving. Ordinarily, with an electric multiple unit where the pick-up shoes were at either end of the train, or a diesel locomotive, these gaps were of no importance as power was being supplied from either end of a train that was much greater in length than any gap, or else the electric power was not being used anyway. With an electric locomotive, this was not the case and it has to be admitted that I was also nearly caught out sometime later.

From the cab of a modern locomotive, the whole railway panorama is set out before you. Modern signalling systems give bright, clear indications regarding the state of the route ahead and modern braking systems give ample power to be able to stop relatively quickly. In the steam era, the view ahead was heavily restricted by the boiler, signals were just a pinprick of candle power – often set against a backdrop of city lights – and braking was often limited to only a portion of the train using vacuum brakes.

The change in footplate environment cannot be overemphasised; it was henceforth clean, warm and easy. The only drawback was that with a steam engine cigarette butts were thrown into the fire but on modern engines the ashtrays needed emptying frequently.

Many of the older drivers were not at all comfortable with the new forms of traction. Steam they knew inside out, every nut and every bolt. They knew exactly what was happening with their engine at any given time and they knew how to judge speed without the need of a speedometer. If something went wrong, they could usually make running repairs to get their train to safety. These new-fangled boxes on wheels were an unknown bit of machinery whirling away behind the bulkhead, of which they took no interest and had no way of effecting repairs if the thing broke down – which was not infrequent.

Consequently, it was quite normal that when a driver was paired with a secondman he could trust, the latter did the driving; nominally under the guidance of the driver. As a result, most of the former fireman who had now become secondmen spent much of their working life actually being unofficial drivers.

No. 34101 *Hartland* departing Southampton Central.

No. 34101 *Hartland* again, ex-works at the rear of Eastleigh shed in number 1 road waiting to be lit up.

It was on one such occasion, when working freight from Southampton Docks to Feltham, that I found myself once again in the driving seat on an electro-diesel locomotive. We exited the non-electrified dock railway system under auxiliary diesel power and dropped the shoes to switch to electric once on the main line. All was well and uneventful until we approached Virginia Water station. This is the location for the junction with the Addlestone line, along which we were travelling, and the main Reading line, which we needed to use for the last few miles to Feltham. It has always been the practice to stop freight trains at Virginia Water before allowing them access across the junction. There was a very long gap in the third rail right across the junction and it was the usual practice to start the auxiliary diesel engine as soon as the train had stopped in the platform. The line was on a sharp incline and a tight curve, which meant that power had to be kept on all across the junction just to keep the train moving. In the steam era, this was always a struggle for men and machine. In the modern era, it was only a struggle for the machinery.

On this occasion, we approached the red colour light signal at the end of the platform slowly and cautiously with light power on against the drag of the incline and the sharp curve. At about 5mph and 20 yards or so from the signal, the red aspect suddenly changed to green. Thinking it was our lucky day in not having the diesel engine started, I immediately notched up the electric power significantly to pull across the junction.

I realised my mistake in time to shut down the electric power enough to minimise the arc of current as our pick-up shoes ran out of third rail. The whole night sky was lit by a bright blue flash, but at least we did not blow the shoes off.

However, we were in trouble. We were foul of the junction with all power gone. The train, which was only travelling slowly up the hill and round the curve to start with, now decided it was time to stop and start to run back again. Naturally, the diesel auxiliary engine also decided it was not going to start very quickly and by the time we were ready to apply power again, we were actually slowly going back the way we had just come. Bringing the train to a stand and restarting with the little diesel engine all takes time and was performed under the not very amused gaze of the signalman.

Little incidents like these helped to inject some interest into what was fast becoming a very boring and unsatisfying job. There was no longer any achievement in a day's work. Morale plummeted and those that were once the elite of railwaymen were now just another grade who worked for the railway.

The year 1967 was not only when steam finished, it was during a time when the whole railway changed and started a downward spiral of cuts and rationalisation. Traffic flows disappeared almost to nothing as freight was switched from rail to road haulage and it became very difficult to see any future prospects of promotion to driver. There was very little left to drive. Radical decisions had to be made and very, very reluctantly I came to the conclusion that my own future lay elsewhere.

My final duty entailed working a petrol train to Kew Junction with a diesel locomotive and working a cement train back to Eastleigh with an electro-diesel. On that November night in 1970 I walked back to the enginemen's mess room, now on Eastleigh station. The place was deserted and dull; all modern and plastic. A few short years earlier the mess room at Eastleigh shed would have been alive with men coming and going, banter, leg pulling and good-natured rivalry. The place had a warmth, character and homeliness that successive generations of enginemen had known, but that had been swept away now, along with those same enginemen.

There were but two men in the office as I signed off for the last time. The duty supervisor was chatting to one of the most senior drivers as I announced this final act. In answer to the inevitable question, I stated that I was happy to be leaving what had become an apology for a railway, that I would no longer have to spend nights away from my bed, and that henceforth I would just forget about the railways.

The senior voice interrupted my flow of bravo. 'No you won't son, you're an engineman as much as the rest of us and you'll always be an engineman.' Dismissing this information with a disparaging comment, I left.

In many ways I am pleased I did not have to be part of the total collapse of the railway system during the 1970s, '80s and early '90s and I was able to go back to finish my interrupted education and move on to pastures new.

And yet, whatever I have done successively, that old driver was quite right. At heart I will always be a steam engineman.

Southampton Central platform 4 with No. 34057 *Biggin Hill* on the daily Brighton to Bournemouth service.

Index

Dedication

And finally to finish this work, a dedication to the memory of all those many drivers and fireman who served at Eastleigh Motive Power Depot from its inception in 1903 until its demise in 1967.

Especially to Drivers P. Bevan and D. Batten – because they had to work with me.

The

Southern Way

The regular volume for the Southern devotee

MOST RECENT BACK ISSUES

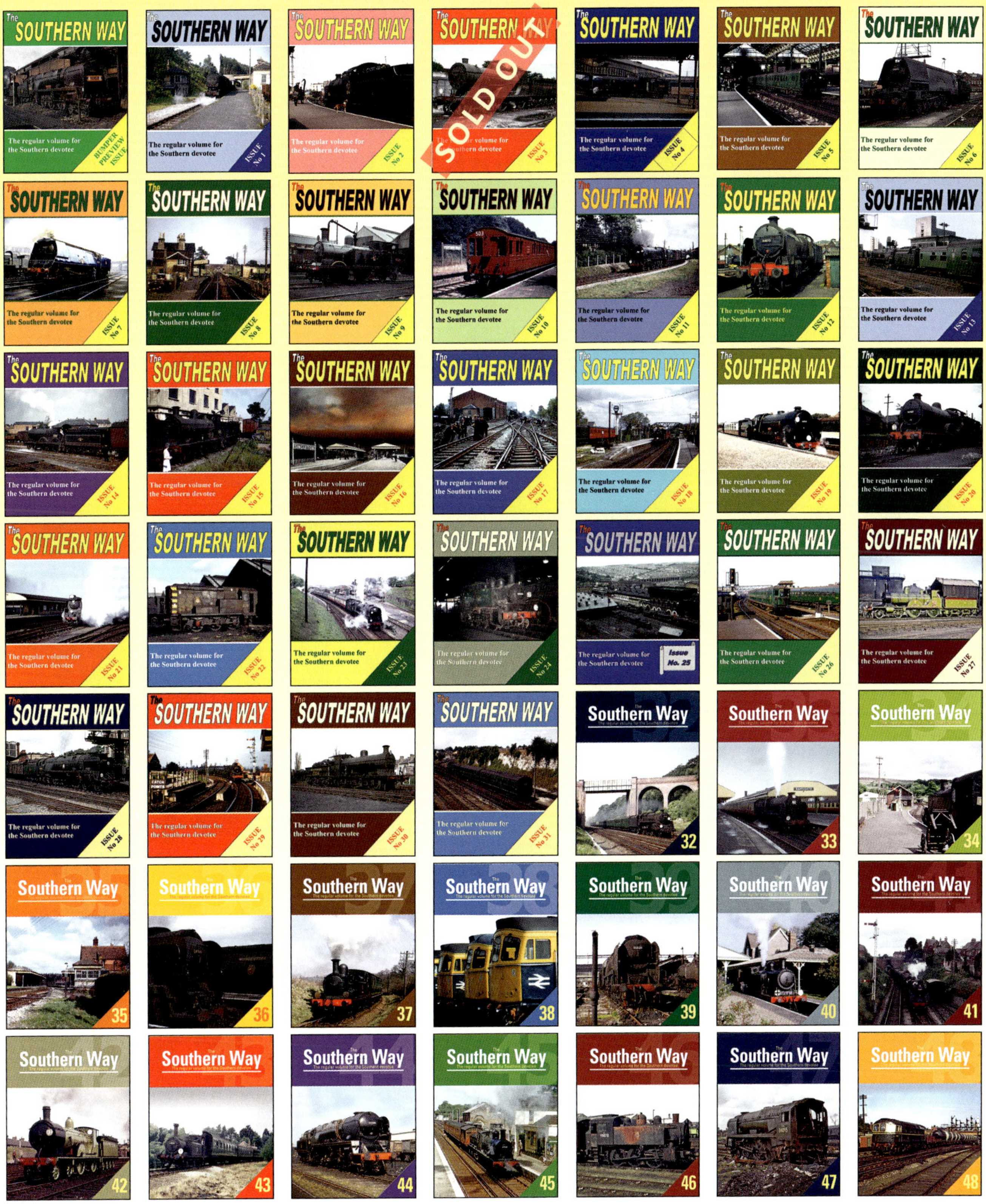

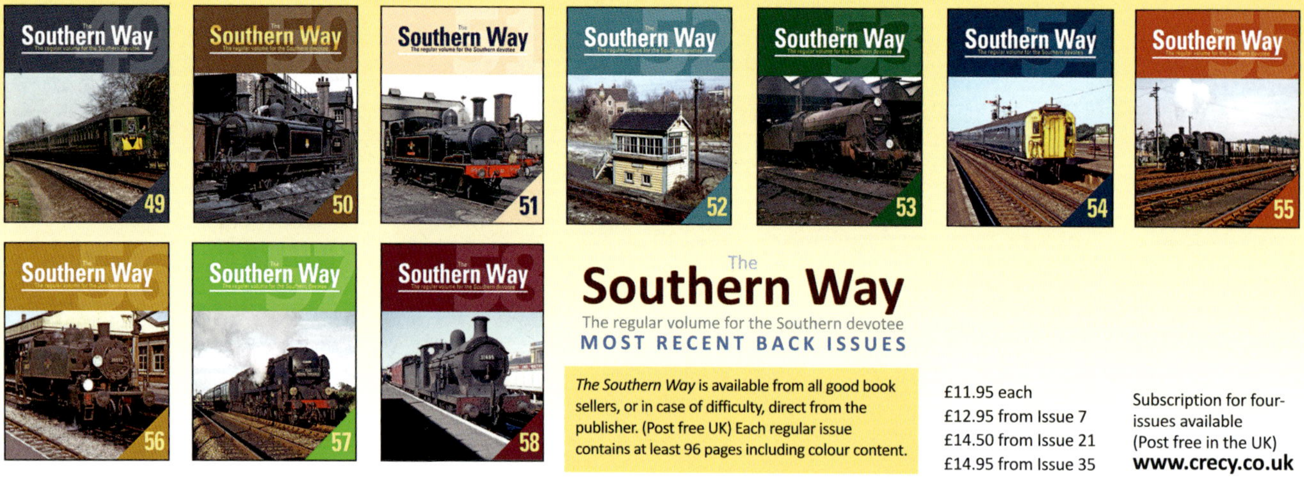

The

Southern Way

The regular volume for the Southern devotee

MOST RECENT BACK ISSUES

The Southern Way is available from all good book sellers, or in case of difficulty, direct from the publisher. (Post free UK) Each regular issue contains at least 96 pages including colour content.

£11.95 each
£12.95 from Issue 7
£14.50 from Issue 21
£14.95 from Issue 35

Subscription for four-issues available
(Post free in the UK)
www.crecy.co.uk

The

Southern Way

The regular volume for the Southern devotee

SPECIAL ISSUES

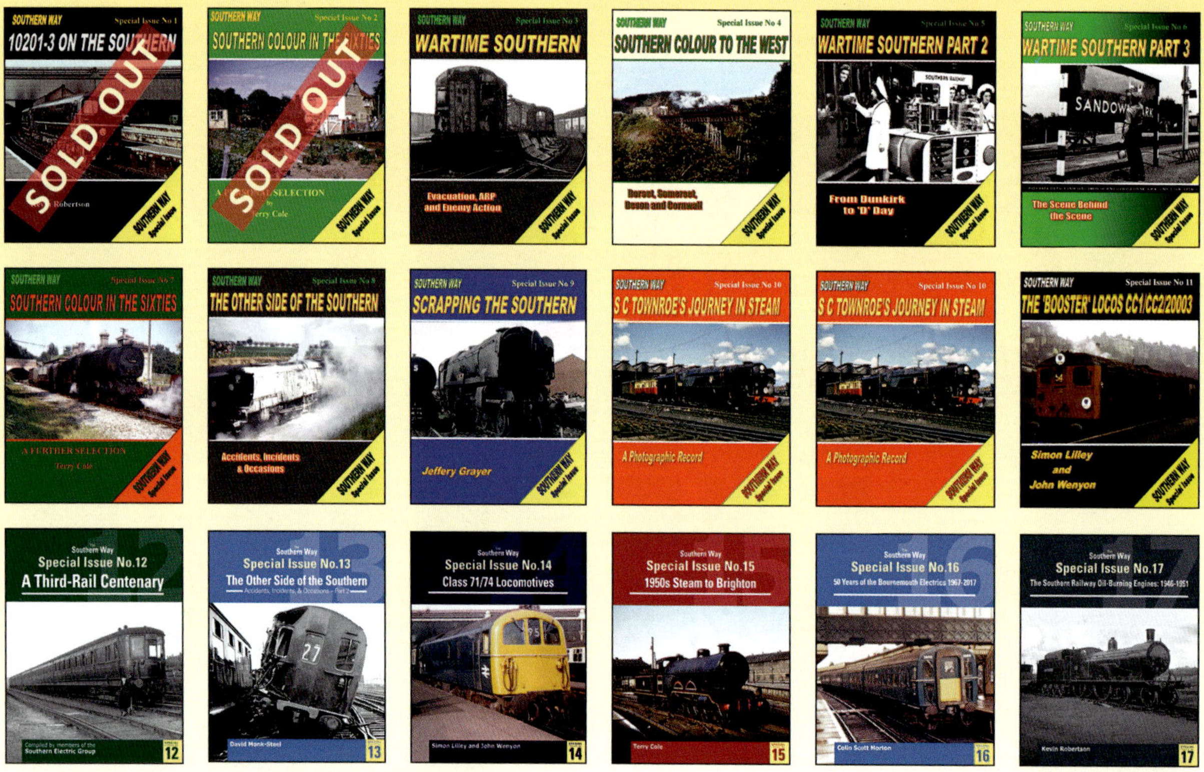

SOUTHERN WAY
'Special Issues'

Nos 3 to 5 £14.95. Nos 6 to 12 £16.50. No 13 on £16.95

Orders, subscriptions and sales enquiries to:

Crécy Publishing
1a Ringway Trading Est,
Shadowmoss Rd
Manchester M22 5LH
0161 499 0024
www.crecy.co.uk